Climate Change from a Criminological Perspective

Rob White
Editor

Climate Change from a Criminological Perspective

Springer

Editor
Rob White
School of Sociology and Social Work
University of Tasmania
Tasmania, Australia

ISBN 978-1-4614-3639-3 ISBN 978-1-4614-3640-9 (eBook)
DOI 10.1007/978-1-4614-3640-9
Springer New York Heidelberg Dordrecht London

Library of Congress Control Number: 2012938717

Printed on acid-free paper

Springer is part of Springer Science+Business Media (www.springer.com)

Contents

Contributors

Robert Agnew Department of Sociology, Emory University, Atlanta, GA, USA

Avi Brisman Department of Criminal Justice, Eastern Kentucky University, Richmond, KY, USA

Andrew Franz Department of Criminology, University of South Florida, Tampa, FL, USA

Pete Fussey Department of Sociology, University of Essex, Colchester, UK

Diane Heckenberg School of Sociology and Social Work, University of Tasmania, Hobart, TAS, Australia

Ingrid Johnston School of Sociology and Social Work, University of Tasmania, Hobart, TAS, Australia

Ronald C. Kramer Department of Sociology, Western Michigan University, Kalamazoo, MI, USA

James Lehane Holt, ACT, Australia

Michael J. Lynch Department of Criminology, University of South Florida, Tampa, FL, USA

Raymond J. Michalowski Department of Criminal Justice, Northern Arizona University, Flagstaff, AZ, USA

Caroline C. Nobo Charlestown, MA, USA

Rebecca D. Pfeffer Boston, MA, USA

Grant Pink Latham, ACT, Australia

Ragnhild Sollund Department of Criminology and Sociology of Law, University of Oslo, Oslo, Norway

Nigel South Department of Sociology, University of Essex, Colchester, UK

Paul Stretesky School of Public Affairs, University of Colorado - Denver, Denver, CO, USA

Noriyoshi Takemura Department of Law, Toin University of Yokohama, Yokohama, Kanagawa, Japan

Rob White School of Sociology and Social Work, University of Tasmania, Hobart, Tasmania, Australia

Chapter 1
The Criminology of Climate Change

Rob White

Introduction

Life on planet earth is rapidly being transformed in ways that are profound and for many of us unimaginable. This is because of global warming that is causing massive shifts in climate worldwide. Until very recently (see White 2009; Stretesky and Lynch 2009; Lynch and Stretesky 2010), few criminologists have participated in analysis of climate change, criminology and criminal justice. This is starting to change and undoubtedly more criminological research and scholarship will be undertaken in this area in the coming years.

As a contribution to needed work in this area, this book stands as the first of its kind to explicitly and specifically deal with climate change from a criminological perspective. Each chapter deals with a substantive issue relating to crime and climate change, and together they forge a disturbing picture of what is happening and what will happen as a consequence of the huge climate-related changes now apparent around the globe. Before outlining the specific contribution of the various chapters, a few preliminary comments help to frame the issues in general posed by climate change and the dilemmas that many of us will face sooner than later in the next few years.

The Slow Crisis of Climate Change

The most pressing issue facing the world today is indeed that of climate change. It is a problem that is global in nature, yet its impacts will be felt at the local and regional levels. Climate change is affecting everyone on the planet, even though

R. White (✉)
School of Sociology and Social Work, University of Tasmania,
Private Bag 17, Hobart, Tasmania 7001, Australia
e-mail: r.d.white@utas.edu.au

R. White (ed.), Climate Change from a Criminological Perspective,
DOI 10.1007/978-1-4614-3640-9_1, © Springer Science+Business Media New York 2012

certain population groups are more adversely affected than others (White 2011). The divides between North and South, geographically and metaphorically, are already deepening as crises related to food production and distribution, energy sources and pollution, and changing climates reorder the Old World order. Social inequality and environmental injustice will undoubtedly be the drivers of continuous conflict for many years to come, as the most dispossessed and marginalized of the world's population suffer the brunt of food shortages, undrinkable water, climate-induced migration and general hardship in their day-to-day lives. Women will suffer more than men, people of colour more than the non-indigenous and the non-migrant, the young and the elderly more than the adult, and the infirm and disabled of all ages.

In many ways, and from the vantage point of future generations, present action and lack of action around climate change will most likely constitute the gravest of transnational environmental crimes. These harms grow more evident every day, yet the main protagonists continue to support policies and practices that contribute to the overall problem (Bulkeley and Newell 2010). Even with foreknowledge and scientific proof in hand, powerful interests continue to dominate the climate change agenda to the advantage of their own sectional interests (see Chap. 4), and it is the poorest of the poor who currently experience the harbingers of things to come for the rest of us (Shiva 2008). Failure to act, now, is criminal. Yet, things continue much as they have, the status quo is maintained, and the harms mount up.

Global warming describes the rising of the earth's temperature over a relatively short time span. *Climate change* describes the interrelated effects of this rise in temperature: from changing sea levels and changing ocean currents, through to the impacts of temperature change on local environments that affect the endemic flora and fauna in varying ways (for instance, the death of coral due to temperature rises in sea water or the changed migration patterns of birds). *Weather* is the name we give to the direct local experience of things such as sunshine, wind, rain, snow and the general disposition of the elements. It is about the short-term and personal, not the long-term patterns associated with climate in general. As the planet warms up, the climate will change in ways that disrupt previous weather patterns, and will in some places even bring colder weather, although overall temperatures are on the rise (Lever-Tracy 2011).

The urgency and reality of climate change issues was eloquently conveyed in a May 2010 letter signed by 284 members of the US National Academy of Sciences published in *Science* claiming that:

> There is compelling, comprehensive and consistent objective evidence that humans are changing the climate in ways that threaten our societies and the ecosystems on which we depend. Many recent assaults on climate science and, more disturbingly, on climate scientists by climate change deniers are typically driven by special interests or dogma, not by an honest effort to provide an alternative theory that credibly satisfies the evidence … (quoted in Lever-Tracy 2011, pp. 10–11).

If anything, most disagreement surrounding climate change today is over how quickly global warming is proceeding rather than over whether it is happening. The majority of governments now acknowledge that climate change must in some way be addressed although concerted action at all levels of government including the international has been decidedly slow and inadequate.

Our understandings and responses to the heating up of the planet fuel disagreement—scientific and political—when it comes to the causes of climate change. These understandings also carry with them certain expectations about potential courses of action. How this question is answered has important social and economic consequences now and in the future (White 2011). For example, if climate change is "natural", then what governments can do is to try to adapt to changed circumstances as best as they can, since change is inevitable (and blameless). In this scenario, existing institutions are not perceived to be the cause of the problem, although they will nonetheless be implicated in the changes that must occur into the future (such as limits being put on carbon emissions).

On the other hand, if human activity is found to be at the genesis of climate change, then this implies that substantial change is needed to the dominant mode of production. Global production and consumption patterns for example feature an insatiable energy appetite (which, in turn, justifies use of destructive energy sources such as coal-fired power stations) and are founded upon a growth model (that feeds polluting and waste industries). The attribution of global warming to human activity also assigns a certain responsibility to the most polluting and damaging industries and countries to make right the wrongs to which they have contributed through their actions. Mitigation and adaptation in this scenario demand redress as well as a major alteration in existing ways of doing things.

Part of the reason why responses to climate change have been so little and so late has to do with the nature of "slow crisis". Floods in Brazil, Australia and Sri Lanka in early 2011 have generally been interpreted publicly as once-in-a-hundred year phenomena. Cyclones and hurricanes are "normal" to certain regions of the world, even though the frequency and intensity might be changing. There is no one single earth-shattering event that demarcates the "crisis" of climate change. Transformation is progressive and longitudinal. It is not abrupt, completed or singularly global in impact.

Yet the consequences of global warming are already apparent. One of these is an expected upsurge in social conflict. Social conflicts are essentially disagreements between different sets of people, and between different nation-states. Four trends can be identified where climate change and associated environmental transformations are giving rise to significant social conflict (see White 2009). These include: conflicts over environmental resources (e.g. access to water, preservation of forests); conflicts linked to global warming (e.g. climate-induced migration); conflicts over the differential exploitation of resources (e.g. foreign patents and imposition of genetically modified crops versus traditional knowledge and use of plants); and conflicts over the transference of harm (e.g. cross-border and global concentrations of pollutants that contribute to global warming).

In one sense, borders do not have much material relevance when it comes to environmental harm associated with global warming. Climate change affects us all regardless of where we live and regardless of social characteristics. But the effects and impacts of climate change differ according to how different individuals and groups are positioned in regard to the global political economy.

For instance, when subsistence fishing, farming and hunting withers due to overexploitation and climate change, then great shifts in human populations and in resource use will take place (White 2011). The forced migration of environmental refugees poses a whole new set of questions for public policy and social justice (see for example, Refugee Studies Centre 2008). Indeed, the relationship between environmental change, climate-induced displacement and human migration is already generating much angst within some Western government circles (Solano and Ferrero-Waldner 2008) and is reinforcing the development of a fortress mentality within particular jurisdictions (whether this is the joined-up countries such as the European Union or discreet nation-states such as Australia).

The global ecological situation is likely to get much worse before too long, particularly as the Arctic heats up. The damage will be felt in the form of extreme weather events, increased competition for dwindling natural resources, outbreaks of disease and viral infections, further extinctions of species, continued pressure to trade off food for fuel, and the list goes on.

The urgency of and need for progressive criminological intervention is illustrated by this kind of charting up emergent social issues related to climate change (see Chap. 2). Managing social conflict, much less dealing with the grossest incidences of environmental harm, will demand great resolve, sharpened analytical tools and high level strategic thinking. It also demands that we interrogate the causes of specific conflicts, the general deterioration of global environmental systems and the distributions of power, energy and wealth on a world scale. There is much work that needs to be done.

Criminality and Climate Change

A number of specific criminal and environmental offences are linked to the phenomenon of climate change. As the consequences of global warming manifest in significant climate changes, there will be various associated offences, some of which will be seriously criminal in nature. Examples of these are presented in Fig. 1.1. The categorization of offences is based upon offences that contribute to climate change, those arising from its consequences, and those pertaining to regulation and law enforcement associated with mitigation and adaptation strategies.

Public policy and governmental action on climate change ultimately needs to address the causal foundations of global warming rather than attempting to just deal with managing the symptoms (Lynch and Stretesky 2010). If the latter course of action is taken, then the tendency will be toward controlling people and repression of conflict, rather than empowering people and getting to the nub of the underlying problem.

Fundamentally, criminology needs to critically examine the consequences of global warming for national security, societal peace, and social and ecological well-being. As part of this engagement with climate change, future research will need to

Environmental Offences (contributing to climate change)

Subject of Offence	**Nature of Offence**
Forestry	illegal felling of trees
Air pollution	emissions of dark smoke
Industrial pollution	unlicensed pollution
Illegal land clearance	destruction of habitat and forests
Clearing native vegetation	reducing biotic mass

Environmental Offences (consequences of climate change)

Subject of Offence	**Nature of Offence**
Water theft	stealing water
Wildlife poaching	illegal killing of animals
Illegal fishing	diminishment of fish stocks

Associated Offences (civil unrest and criminal activities)

Subject of Offence	**Nature of Offence**
Public order offences	food riots
Eco-terrorism	arson, tree spiking
Trafficking	migration and people smuggling
Violent offences	homicide, gang warfare

Regulatory Offences (arising from policy responses to climate change)

Subject of Offence	**Nature of Offence**
Carbon trading	fraud
Carbon offsets	misreporting
Illegal planting	unauthorised use genetically modified organisms
Collusion	regulatory corruption

Fig. 1.1 Crime and climate change: Offences

be multi-jurisdictional in scope and transnational in nature. Key issues include the following:

- *Bio-security, national security and conflict resolution* (e.g. in relation to specific environmental crimes related to food production/consumption such as illegal fishing; issues such as climate-induced migration and border security; issues pertaining to disasters, crime and the breakdown of law and order; the application of situational and other crime prevention techniques in a proactive manner around environmental issues)
- *Climate change and environmental enforcement and regulation processes* (e.g. in relation to carbon emissions, and to carbon emission trading and reduction schemes, and implications for environmental law enforcement agency practices; sustainable development strategies and compliance strategies in regard to new environmental regulations and standards in areas such as water and energy use)
- *Climate change strategies, land use and waste disposal* (e.g. issues arising from specific strategies designed to mitigate/adapt to climate change such as mass adoption of mercury-filled compact fluorescent light globes; the production, transportation and storage of wastes such as radioactive waste; the impact of windmills as energy generators in relation to specific land uses and threats to particular species)

How we interpret and respond to global developments such as climate change depends upon how we define environmental harm, how we envisage the protection of human, ecological and animal rights, and how we understand the social and political relationships that underpin recent trends and issues. For criminology, there is no doubt that new typologies of harm have to be developed, new methodologies for global research instigated, and new modes of social control devised is we are to adequately address climate change.

For example, in analysis of climate change issues a variety of concepts will need to be deployed in investigating substantive matters. Certainly time, space and scale are relevant to such analysis. For example, risks and harms may be direct or indirect, and their consequences may be felt in the immediate or in the long-term. Harm may be specific to local areas (such as threats to certain species, like coral in the Great Barrier Reef) yet manifest as part of a general global pattern (such as being an effect of wide scale temperature changes affecting coral everywhere). Harm is central, but this may be non-intentional (in the sense of being a byproduct of some other agenda) or premeditated (insofar as the negative outcome, for some, is foreseen). The demise of the polar bear due to the impact of global warming in the Arctic is an example of the former. The displacement of local inhabitants from their land due to carbon sequestration schemes is an example of the latter.

Some key emergent or horizon issues of relevance to criminological study of climate change stem from the fact that collective security will increasingly be tied up with notions of ecological sustainability within a particular social context (White and Heckenberg 2011). Pressures relating to food and water supply, and loss of habitat, will manifest in various class-related processes including certain types of criminality (see also Smith and Vivekananda 2007). For instance, class-related "future" crime will include: *crimes of the less powerful* (e.g. theft and violence as responses to survival needs, such as loss of land, food shortages, loss of livelihoods and escalating poverty) and *crimes of the powerful* (e.g. profit motive related to climate change, such as take-over of land in relation to carbon emission trading schemes and transformation of food crops into biofuel crops regardless of local subsistence needs). Scarcity and conflict over natural resources will be linked to corruption, illicit markets, riots and profiteering. The forced movement of people will be tied up with exploitation such as trafficking, child soldiers and sexual slavery.

Simultaneously, climate change will accelerate the extent and intensity of so-called "natural" disasters such as cyclones, tsunami and mudslides (White and Heckenberg 2011). Issues pertaining to disaster studies include investigation of crime related to the disaster, such as looting, rape, and the general breakdown of law and order. Other crimes relate to pre-conditions, such as the contribution of poor building codes to fatalities, and to criminality such as fraud and contractor misdealing when it comes to rebuilding after a disaster has occurred. Questions can be asked about social responsibility and causal chains in relation to certain trends and events—for example, climate extremes due to production based upon carbon emission technologies, or mudslides that ultimately stem from deforestation on the top of mountains and hills. There is evidence that women and children are especially prone to victimization in disaster settings and that they suffer disproportionately (relative to men, but also in relation to ordinary crime occurrences) when it comes to sexual assault and family violence (Thornton and Voigt 2007; Chap. 9). Frustration at lack of sufficient aid can also lead to post-disaster riots, anger at authority figures such as the police and the formation of gangs.

Climate Change, Crime and Criminology

This book provides detailed and reflective analyses of just these kinds of issues. Be this as it may, it is nonetheless just the beginning of the kind of work that is needed in this area. Time is short, there is much to do, and yet criminology has lots to offer, too, in understanding and proposing action around climate change issues.

The next three chapters chart out the general criminological terrain for the study of climate change. The chapter by Agnew draws upon major criminological theories in order to make a rather disturbing argument: namely, that the advance of climate change will reduce rather than increase the likelihood of meaningful action. The chapter provides brief overviews of the effects of climate change, the inadequate response to such change, and the reasons for this inadequate response. It then discusses consequences of climate change from a criminological perspective, consequences such as increased strain, reduced control and greater social conflict. These consequences, in turn, are said to reduce the ability and willingness of individuals and groups to take meaningful action on climate change.

In a similar vein, Fussey and South chart out the coming "climate divide" that represents a further extension of the contemporary inequitable state of the affairs of humanity and the planet. It is argued that not only will those with the fewest resources have the greatest difficulties in mediating the impact of climate change and its attendant shocks, but climate change will stimulate a number of deeply criminogenic forces. Together, such interconnectivity between the global and local suggests that approaches to sustainability and resilience need to be broadly conceived in both scope and application and need to be genuinely transformative rather than operating within current ambitions for "business as usual". Moreover, the magnitude of these issues underlines the importance of formulating an approach to sustainability and resilience that genuinely embeds the "green" of environmental concerns within the "blue" of security policy.

Brisman presents a view of climate change from the perspective of cultural green criminology. The substantive focus of this chapter is to detail how climate change contrarianism is manifested in the mass media. Denial is profoundly ideological in nature; how it is conveyed and transmitted is of importance to those who wish to reorient collective thinking to not only recognizing the urgency and seriousness of the problem, but recasting it in criminal terms. It is vital to understand and expose the dynamics and social construction of deception and "contrary" opinion if positive action is to be taken to address climate change issues.

The next four chapters deal with aspects of state involvement in climate change related issues. Kramer and Michalowski pose the question of whether global warming is a state-corporate crime. This chapter examines how transnational corporations and the nation states of the global North act in concert in ways that cause widespread environmental and social harm by, for example, denying that global warming is caused by human activity and blocking efforts to mitigate greenhouse gas emissions. The chapter concludes by arguing that criminologists need to engage in a public criminology that communicates the relationship between state-corporate crimes and environmental degradation to audiences beyond their academic peers.

The chapter by Franz focuses on the harms of climate change and the response to those harms from a legal perspective. Recurring and emerging issues in American jurisprudence of climate change (standing, legal harm, political question doctrine, agency capture, and evidence laws regarding science) are discussed and critiqued. The legal response to climate change through international (regional, European Union and United Nations) agreements, developments and doctrines, is also considered as well as the use of relevant legal tribunals outside of the USA to address such harms. Overall positive trends in international law and customary law are noted, as well as certain practical limitations of emerging doctrines and practices.

The role of environmental enforcement networks forms the substantial topic of the chapter by Pink and Lehane. Climate change regulation, like all forms of regulation, requires allegations of non-compliance to be investigated. Enforcing climate change law and regulations is already sufficiently challenging given a myriad of social, economic and environmental issues. However, climate change regulation is further complicated due to cross-jurisdictional issues, transnational factors and its intersection with traditional and cross-over crimes such as fraud and money laundering. It is anticipated that both non-compliant entities and organized criminal enterprises will challenge and frustrate the efforts of government regulators as they attempt to enforce climate change legislation. This chapter explores the experiences of enforcement and regulatory agencies that have cooperated and worked collaboratively through various Environmental Enforcement Networks to advance and maximize their enforcement effort.

The chapter by Sollund focuses on the inconsistency of Norwegian policies which on the one hand produce severe damage to the environment and to the climate (related to its oil production), yet on the other hand attempt to modify the damage through specific climate related projects, such as compensating for its carbon emissions by investing in rain forests in Brazil and Indonesia to prevent deforestation. On a local level climate change is apparent in the Norwegian mountains where the polar fox species which feed on lemmings which again depend on snow for their survival has been brought near to extinction. To prevent the disappearance of the species, a programme to save the polar fox was initiated through which polar fox cubs are bred and released in the mountains. Generally, the mortality rate has been huge and animals have suffered from starvation. The discussion then turns to individual rights versus species rights and justice, and the moral right to individually abuse individuals to secure species survival. This discussion is informed by the treatment of polar foxes which were "contaminated" with the wrong genes and consequently killed.

The next few chapters consider the relationship between climate change and various types of disaster. Heckenberg and Johnston examine the issue of climate change, extreme events and natural disaster and the way in which gender relations place men and women differently at risk depending on the type and pace of an event and where it occurs. The chapter explores some of the experiences of males and females, and of adults, youth and children in response to different events across the world. In the process, questions arise about the role of gender in shaping the perspectives, vulnerabilities and responses of different groups. The chapter concludes with a proposal to draw on the principles of crime prevention (factors influencing situational and opportunistic crimes) and the practice of horizon scanning (extrapolating what is currently known about the gendered landscape of climate change and disaster to future events) to formulate contingency plans to mitigate potential environmental degradation, to reduce crime and criminality during disasters and to avert gendered victimization.

The chapter by Nobo and Ofeffer provides discussion of the criminological lessons to be learned from analysis of Hurricane Katrina. It is observed that one of the most severe consequences of global climate change is an increase in the intensity and frequency of extreme weather events. An unanticipated natural disaster can devastate an area with physical damages. In addition, it presents law enforcement with an unprecedented moral and organizational challenge, especially when ill-equipped criminal justice personnel are thrust into a situation without the resources to effectively police and regulate a post-disaster environment. Criminal justice malfeasance following Hurricane Katrina demonstrates a paradox in which those charged with upholding their oath to protect and serve instead faltered and proved unable to provide a safe infrastructure for the citizens of New Orleans. What Hurricane Katrina taught us is that extreme weather yields extreme human reactions.

The chapter by Takemura considers the big earthquake and tsunami on the 11 March 2011 that triggered a major nuclear power plant accident in Japan, which has caused and will continue to cause tremendous damage and harm. This "accident", however, is not seen as a natural disaster but as human-made disaster, and as one of the most serious crimes committed by the state–corporate complex in Japan's history. From the point of view of climate change, we can anticipate many more natural disasters such as tsunami and cyclones as the earth warms up and new climatic conditions emerge. In the midst of more frequent and more intense climate-related events, it is essential that people everywhere be prepared for the potential advent and devastating aftermath of profound natural disasters of this kind. Yet, as this chapter demonstrates, certain vested interests in the corporate sphere and the state arena are already stifling adequate knowledge, discussion and action around these sorts of questions. Insofar as this remains the case, substantive measures to mitigate and adapt to climate change will accordingly be diminished.

The final chapter by Lynch and Stretesky considers the possibility of state action that could be taken now in order to reduce carbon emissions. The main thrust of this chapter is how strategic forms of environmental regulation can make a significant difference in environmental outcomes. The chapter discusses how in the late 1970s, the Corporate Average Fuel Economy (CAFE) standards were created to reduce US dependency on foreign oil. Recently, the purpose of CAFE has been modified to help reduce carbon dioxide (CO_2) emissions among the US automobile fleet. This chapter examines the drawbacks of CAFE to reduce CO_2 emissions and proposes a vehicle based carbon tax (V-CART) that offsets emissions with using carbon credits. The benefits and limitations of this alternative policy are discussed.

Taken together, the chapters in this book provide multiple insights into the huge problems and challenges posed by climate change, and the potential course of action that can be put into place now to address present and emergent impacts related to climate change. They demonstrate the value and contributions that criminology as a field can make in dealing with climate change issues. However, as reinforced throughout this text, it is even more important—and urgent—that further work be undertaken in this area. Our lives and future depend on it.

References

Bulkeley, H., & Newell, P. (2010). *Governing climate change*. London: Routledge.

Lever-Tracy, C. (2011). *Confronting climate change*. London: Routledge.

Lynch, M., & Stretesky, P. (2010). Global warming, global crime: a green criminological perspective. In R. White (Ed.), *Global environmental harm: Criminological perspectives*. Devon: Willan Publishing.

Refugee Studies Centre. (2008). *Forced migration review: Climate change and displacement, Issue 31*. Refugee Studies Centre: University of Oxford.

Shiva, V. (2008). *Soil not oil: Environmental justice in an age of climate crisis*. Brooklyn: South End Press.

Smith, D., & Vivekananda, J. (2007). *A climate of conflict: The links between climate change, peace and war*. London: International Alert.

Solano, J., & Ferrero-Waldner, B. (2008). *Climate change and international security, Paper from the High Representative and the European Commission to the European Council*. Brussels: European Union.

Stretesky, P., & Lynch, M. (2009). A cross-national study of the association between per capita carbon dioxide emissions and exports to the United States. *Social Science Research, 38*, 239–250.

Thornton, W. E., & Voigt, L. (2007). Disaster rape: Vulnerability of women to sexual assaults during hurricane Katrina. *Journal of Public Management and Social Policy, 13*(2), 23–49.

White, R. (2009). Climate change and social conflict: Toward an eco-global research agenda. In K. Kangaspunta & I. Marshall (Eds.), *Eco-crime and justice: Essays on environmental crime*. Turin, Italy: United Nations Interregional Crime Research Institute [UNICRI].

White, R. (2011). *Transnational environmental crime: Toward an eco-global criminology*. London: Routledge.

White, R., & Heckenberg, D. (2011). Environmental horizon scanning and criminological research and practice. *European Journal of Criminal Policy and Research, 17*(2), 87–100.

Chapter 2
It's the End of the World as We Know It: The Advance of Climate Change from a Criminological Perspective

Robert Agnew

Introduction

The future effects of climate change depend heavily on the steps we take to mitigate and adapt to it. We now have a reasonably good idea of how to avert the worst effects (see below), but so far have largely failed to act on this knowledge. Psychologists, economists, political scientists, and sociologists have devoted significant attention to the factors that have inhibited meaningful action, factors such as low environmental concern, the tendency to respond to immediate rather than delayed threats, and the efforts of the fossil fuel industry and its allies to create doubt about climate change Underlying much of this literature is the assumption that action on climate change will increase once the very harmful effects of such change become apparent.

The chapter draws on the major crime theories to make a rather different argument—that the advance of climate change will *reduce* rather than increase the likelihood of meaningful action. I first briefly discuss the effects of climate change, the inadequate response to such change, and reasons for this inadequate response. I then discuss certain of the consequences of climate change from a criminological perspective, consequences such as increased strain, reduced control, and greater social conflict. These consequences are said to reduce the ability and willingness of individuals and groups to take meaningful action on climate change. I conclude by describing an alternative, more hopeful narrative.

R. Agnew (✉)
Department of Sociology, Emory University, Tarbutton Hall,
1555 Dickey Drive, Atlanta, GA 30322, USA
e-mail: bagnew@emory.edu

R. White (ed.), Climate Change from a Criminological Perspective,
DOI 10.1007/978-1-4614-3640-9_2,

Background: Climate Change and the Failure to Respond

Climate Change and Its Effects

Our climate is changing, due largely to the burning of fossil fuels and, to a lesser extent, deforestation. The global mean temperature is increasing; ocean levels are rising; rain is increasing in some areas and decreasing in others; and extreme weather events are becoming more common—including hurricanes/cyclones, heavy downpours, heat waves, and droughts. Unless we take dramatic action in the very near future, climate change will likely proceed to the point where its effects are catastrophic. There will be massive food and freshwater shortages. Hundreds of millions will lose their homes and livelihoods to flooding, extreme weather events, and habitat change (e.g., grasslands turning into desert). There will be large-scale migrations, with many moving to megacities in developing nations. Social conflict will increase, particularly as groups compete over scarce resources. And billions will have their health threatened due to increased malnutrition, air and water pollution, extreme weather events, and the spread of diseases such as malaria and dengue fever. These effects will be greater among those in developing nations, the poor, females, the very young, and the very old—groups who are more vulnerable and/or less able to adapt to climate change (see also Chap. 9). But all will suffer (e.g., Buhaug et al. 2008; Cullen 2010; Global Humanitarian Forum 2009; Henson 2011; Kolmannskog 2008; Lancet and University College London Institute for Global Health Commission 2009; Oxfam 2009).

The Failure to Take Meaningful Action

Even though we are confronted with perhaps the greatest threat to humanity in our history, there has been little serious effort to mitigate and adapt to climate change. This is particularly true of the United States, which has emitted more greenhouse gases than any other nation, and it is the true at the international level (see Cullen 2010; Henson 2011; Lynch et al. 2010; McCright and Dunlap 2010; McKibben 2010). There is a general consensus that the global mean temperature should not increase by more than 2°C if we are to avert the effects of climate change just described. The greenhouse gases *now* in the atmosphere will result in an increase of about 1.5°C. And the rate of growth in carbon emissions is increasing. At present, we are headed toward an increase of 4–5°C by the end of this century, perhaps sooner (Henson 2011; McKibben 2010).

Efforts to deal with climate change must include a dramatic reduction in the use of fossil fuels, achieved by increasing the cost of carbon emissions, heavy investment in alternative energy sources, increased energy efficiency, alternative methods of farming, population control, and changes in lifestyle (e.g., less automobile use, smaller homes, reduced meat consumption). Efforts must also include initiatives to remove and store carbon from the atmosphere, such as reforestation and the production of biochar (a charcoal that stores carbon, as well as enriching the soil). They must include programs to adapt to climate change, such as efforts to protect coastal areas from flooding and implement sustainable methods of farming. And they must include efforts by developed nations, who bear major responsibility for climate change, to help developing nations adapt and develop their economies in a sustainable manner (for overviews, see Global Humanitarian Forum 2009; Gore 2009; Henson 2011; Hertsgaard 2011). Most notably, a dramatic increase in funding is necessary to help developing nations create or move to renewable energy sources, improve energy efficiency, foster sustainable agricultural practices, support population-control programs, and adapt to such threats as flooding, food and freshwater shortages, climate-related health problems, and forced migration.

Reasons for Inaction

Researchers have devoted significant effort to explaining why most individuals and groups have failed to take meaningful action to mitigate and adapt to climate change. It is impossible to fully review this literature here, but a few general themes emerge (for summaries and selected studies, see Diekmann and Preisendorfer 2003; Dietz et al. 2007; Frantz and Mayer 2009; Franzen and Meyer 2010; Gockeritz et al. 2010; Henson 2011; Iwata 2004; Kurz 2002; Lindenberg and Steg 2007; Lynch et al. 2010; McCright and Dunlap 2010; Parks and Roberts 2008; Patchen 2010; Takacs-Santa 2007; York et al. 2003).

In order for action to occur, climate change must first be *defined as a relatively serious problem*. But this is frequently not the case, for a variety of reasons. Among other things, climate change is a problem of a special sort—it is a "slow crisis." (Chap. 1) Its worst consequences will not appear for many years (e.g., a substantial rise in sea level), but people and groups are inclined to respond to immediate rather than delayed threats. Further, there has been a concerted effort on the part of the fossil fuel industry and others to create doubt about climate change and its negative effects. This effort is reflected in corporate advertising and the support provided to organizations and scientists who question climate change. And the mass media often unwittingly support this effort, in the interest of presenting "balanced coverage" of issues (Henson 2011; Lynch et al. 2010; McCright and Dunlap 2010; Takacs-Santa 2007; Chap. 4).

Second, individuals and groups must feel that they have the *ability to take meaningful action*. But many believe that they are unable to do so, given the structural constraints they face, their lack of resources, and/or the enormity of the problem. Many people, for example, have no choice but to drive to work given the design of cities and lack of public transit. Many developing nations lack the financial resources to adapt to climate change (e.g., build levees to protect against rising sea levels, develop industries less vulnerable to climate change).

Third, individuals and groups must feel that they have *an obligation to act*. This obligation is rooted in their normative beliefs (e.g., belief that others engage in environmentally responsible behavior and expect that they do the same), moral beliefs (e.g., belief that environmentally responsible behavior is good), emotions (anger and sadness about environmental problems), and sense of responsibility for the problem. But many do not feel this obligation. The people around them do not engage in environmentally responsible behavior or expect it of others. They do not accord much value to the environment. Instead, they prioritize economic development, viewing the environment as a resource to be exploited for their benefit. Related to this, they are not upset by environmental problems. And/or they do not believe that it is their responsibility to act. For example, they believe that others created the problem and it is their responsibility to act. Or they believe that climate change will be addressed by technological advances (e.g., Hayward 2010; Patchen 2010; Takacs-Santa 2007). Beyond that, some engage in token behaviors, such as occasional recycling, that make them feel that they have fulfilled their obligation to act.

Finally, individuals and groups must believe that it is in their interests to act, with the *perceived benefits of action outweighing the costs*. But many believe that the costs of action are too high (e.g., Diekmann and Preisendorfer 2003; Lindenberg and Steg 2007; Lynch et al. 2010; Patchen 2010). Many individuals, for example, believe that it takes too much time and effort to walk or use public transit. Many corporations believe that action would jeopardize their profits or even survival. Many in the United States believe that action would threaten their economic prosperity, while many in the developing world believe that it would threaten their plans for economic development. A good case can be made that such views are mistaken, particularly when long-term interests are considered, but such views nevertheless contribute to inaction.

Given the above, some argue that there will not be a meaningful response to climate change until the negative consequences of such change become more apparent. At that point, climate change will become an immediate problem for much of the world; its seriousness difficult to deny. Many will feel an obligation to act, given that climate change will become an *obvious* threat to humanity—as well as to economic prosperity and development. And many will feel that it is in their interests to act, given that the costs of climate change will have become quite high. Unfortunately, it will then be too late to avert many of the effects described above. Climate change and its effects cannot be quickly reversed. The greenhouse gases we emit stay in the atmosphere for decades or longer, and after a certain temperature threshold is crossed we may set in motion processes that rapidly increase the rate of climate change. For example, a further rise in temperatures may result in the massive release of methane—a very potent greenhouse gas—from the arctic tundra and ocean floors. This in turn would result in a dramatic increase in climate change, even if the burning of fossil fuels were dramatically reduced (see McKibben 2010, pp. 20–22). Nevertheless, there is some feeling that meaningful action will come with the advance of climate change. These ideas are expressed by Giddens (2009, p. 2), among others:

> Since the dangers posed by global warming aren't tangible, immediate, or visible in the course of day-to-day life, however awesome they appear, many will sit on their hands and do nothing of a concrete nature about them. Yet waiting until they become visible and acute before being stirred to serious action will, by definition, be too late.

The Response to Climate Change: A Criminological Perspective

I next question this view, arguing that the advance of climate change will *reduce* the likelihood of meaningful action to mitigate and adapt to it. I draw on the leading crime theories in making this argument, particularly strain, social support, social control, social learning/rational choice, and critical theories (see Agnew 2009; Cullen and Agnew 2011). These theories suggest that the consequences of climate will be such that they create more pressing problems for individuals, reduce the ability to take meaningful action, undermine the obligation to act, and increase the costs of such action. It is not my intention to be a doomsayer in making these arguments, but rather to point to certain unanticipated consequences of climate change in the hope that we might better respond to them.

Climate Change as a Relatively Serious Problem

Many people now view climate change as a serious problem; even in the United States—where there has been a major campaign to raise doubts about it. In a 2010 Gallup poll, for example, 53% of the people surveyed in the United States agreed that "global warming" was a "very" or "somewhat" serious threat (42% of the people in the 111 countries surveyed said the same). When people are asked to list the *most important* issues facing the United States, however, the environment—including climate change—is typically listed by less than five percent of respondents in open-ended questions. Similar results are obtained in Europe. And the environment is ranked at or near of the bottom when respondents are presented with lists of problems to rank (Hayward 2010; also see Dietz et al. 2007). As a result, climate change is one of many problems competing for resources, with other problems taking priority.

Drawing on strain theory (Agnew 2006), I argue that the advance of climate change will *not* increase the *relative* priority attached to it. This may seem like an odd argument; research suggests that concern about environmental problems is partly a function of vulnerability to, experiences with, and media reports on such problems (Brody et al. 2008; Takacs-Santa 2007). All of these will increase given the massive problems that climate change will bring. Strain theory, however, suggests that while these problems will increase the *absolute* level of concern about climate change, they will not increase the *relative* priority attached to it. To understand the basis for this argument, it is first necessary to describe the problems or strains that will result from climate change. These strains include the following:

- Extreme weather events, including hurricanes, heat waves, droughts, and floods, as well as associated phenomena such as forest fires and blackouts (Henson 2011; Oxfam 2009).
- Massive food, freshwater, and fuel shortages, especially in the developing world. Billions of people, for example, will face extreme freshwater scarcity (Global Humanitarian Forum 2009; Oxfam 2009).
- The loss of or threats to livelihood, particularly on the part of farmers, herders, fishers, and those involved in snow- and water-related tourism (60% percent of the population in developing nations). Partly as a result of such threats, poverty and inequality will increase (the poor will suffer proportionately more than the rich from climate change, thus increasing levels of inequality) (Global Humanitarian Forum 2009; Oxfam 2009).
- The loss of homes and property due to extreme weather events and rising sea levels (Global Humanitarian Forum 2009; Oxfam 2009). For example, sea level is predicted to rise by at least 1 m this century, and perhaps many meters beyond that. Thirteen of the world's 20 largest cities are on the coast, and so are directly threatened by this rise, as well as by the higher storm surges associated with more severe storms.
- Illness and injury due to the spread of disease, food and freshwater shortages, increased air and water pollution, and extreme weather events; as well as the death and injury of close others. The Global Humanitarian Forum (2009) estimates that 300,000 people *now* die each year as a result of Climate Change, and a Lancet Commission report states that "climate change is the biggest global health threat of the twenty-first century," with the lives and well-being of billions of people at risk (Lancet and University College London Institute for Global Health Commission 2009; also see Oxfam 2009).
- Forced migration, both within and across borders, with many moving to the slum areas of megacities in developing nations. Residence in these slums involves exposure to additional strains. Among other things, work and resources are scarce; living conditions are often crowded, noisy, and chaotic; those in the receiving population are sometimes hostile; and criminal victimization is frequently high. Estimates vary a good deal, but many claim that hundreds of millions of people will be forced to move this century due to such things as rising sea levels, desertification, extreme weather events, and social conflict (Kolmannskog 2008; Oxfam 2009; Tacoli 2009).
- Exposure to violent social conflicts and crime, fueled in part by competition over scarce resources and the above strains (Agnew 2011; Buhaug et al. 2008; Kolmannskog 2008; Raleigh et al. 2008; Oxfam 2009; Smith and Vivekananda 2007). In particular, climate change may increase conflict between states (e.g., interstate conflicts over freshwater sources), between groups within states (e.g., conflicts over food, the distribution of disaster relief), and between individuals (interpersonal violence

and theft). Agnew (2011) argues that climate change will become one of the major, if not the major, forces driving crime as the century progresses.

- Strains affecting higher-class individuals, corporations, and developed nations. While the poor will suffer more, wealthier individuals and groups will experience many of the above strains, such as extreme weather events. Also, they will pay more for things such as energy, consumer goods, and insurance. Their lives will be more closely regulated, as efforts to limit carbon emissions increase. And the poor will often make demands on and threaten them, since they are largely responsible for climate change and possess valued resources.

These strains pose an immediate and direct threat to individuals and groups, and so will be ranked highest in priority. In particular, individuals and groups will focus on ways to reduce and escape from these strains, as well as seek revenge against those blamed for them. One might argue that these strains are caused by climate change and so the major focus will be on ways to alleviate and adapt to such change. But when presented with strains of the above type, the immediate focus is on the strains themselves and not on the background factors that cause them. A hungry person, for example, searches for food rather than more sustainable methods of farming. We see evidence for this today, particularly in the United States. When the economic crisis struck in 2008, there was a substantial shift in public opinion, with people becoming much more likely to state that economic growth should take priority over environmental protection (53% felt this way in 2010, versus 23% in 2000) (Hayward 2010). More generally, data suggest that poorer nations and the poorer people within nations have lower levels of environmental concern (Franzen and Meyer 2010).

In addition, climate change is a problem of a special sort. Just as the negative effects of climate change take years to emerge, efforts to mitigate and adapt to climate change take years to have an effect. Efforts at mitigation will take decades or longer, partly because of the extended lifecycle of most greenhouse gases. And most efforts at adaptation will take much time to implement, assuming that the resources for implementation are available. Individuals and groups experiencing the above strains, however, will be in the market for immediate solutions. Further, they will be less selective about the nature of such solutions than would ordinarily be the case. Their desperate state promotes a focus on immediate self-interest, with less concern for the long-term consequences of their behavior and its effect on others. Consequently, they will sometimes cope by committing criminal or harmful acts, including acts that contribute to further climate change. For example, they may steal food, raid forests for fuel, burn low-grade coal, and attack migrant groups. The likelihood of such criminal or harmful coping is further exacerbated by other of the effects of climate change, described below.

The Ability to Take Meaningful Action

Climate change will also undermine the ability of individuals and groups to take meaningful action to mitigate and adapt to it. First, climate change will reduce the resources necessary for such action. As climate change proceeds, the strains described above will become more widespread, frequent, and severe. Further, they will often co-occur; for example, certain areas will experience much increased temperatures, coastal flooding, extreme weather events, *and* drought. As a consequence, levels of poverty will increase, with large numbers of people losing their livelihoods and property. Likewise, large numbers of people will have their physical and mental health impaired, with extreme weather events and other of the above strains increasing depression, lethargy, and stress disorders (Page and Howard 2010). Many will lose their social supports as well. Family members, friends, and neighbors will be killed and injured, experience health problems, lose their resources, and/or move away. Private insurance will become too costly or unavailable. And many will migrate to the slums of megacities, where social support is low. Likewise, communities and states will exhaust their resources as they struggle to cope with the negative effects of climate change. Such effects include food, freshwater, and fuel shortages; widespread death, injury, and illness; the destruction of critical infrastructure by extreme weather events; the displacement of large numbers of people due to flooding, drought, and extreme weather events; and social conflict. As a consequence, individuals and groups will lack the resources—including material, physical, psychological, and social resources—to take meaningful action. This will be especially true for poor individuals and developing nations.

Second, climate change will reduce the ability of groups to exercise social control, also critical if meaningful action is to occur. An effective response to climate change requires action that is not in the immediate interests of individuals and groups. In particular, it requires prohibiting individuals and groups from engaging in certain self-interested acts that contribute to climate change (e.g., excessive driving, building coal-fired power plants). And it requires mandating them to engage in acts that do not serve their immediate interests (e.g., taking public transit, installing emissions controls). That is, it requires that they invest their resources in mitigation and adaptation projects that will not provide benefits for years or even decades. Further, it requires that developed nations provide massive aid to developing nations, to help them adapt to climate change and build sustainable economies. Accomplishing these things requires the exercise of social control, since individuals and groups are often reluctant to act against their immediate self-interests—especially in the face of the threats just described.

Effective social control involves setting clear rules, monitoring behavior, and consistently sanctioning rule violations in a meaningful way. It involves socializing individuals and groups so that they accept the need for such rules. And it involves providing individuals and groups with a stake in conformity, so they have some incentive to abide by the rules (see Agnew 2009). Climate change, however, will reduce the ability of communities, nations, and international organizations to exercise such control. These groups will have to devote more of their resources to coping with the negative effects of climate change just described, and so will have fewer resources to devote to their criminal justice systems and other organizations designed to exercise social control. Extreme weather events and associated phenomena, such as blackouts and forest fires, will reduce the ability to exercise social control by disrupting routine activities. Likewise, the growth of megacities will undermine efforts at control, since it is more difficult to exercise control in very poor areas characterized by high turnover (Agnew 2009). But most importantly, these groups will have trouble meeting the basic needs of people, such as the needs for food and water, shelter, security, and work. As a consequence, people will be less likely to accept the rules promoted by these groups—both because they have a lower stake in conformity and are more likely to question the legitimacy of these groups.

Finally, the advance of climate change will increase both intra- and inter-state conflict, making it difficult to forge the cooperative agreements necessary for a meaningful response to climate change. The advance of climate change will intensify existing conflicts. This includes conflict between those who benefit from activities linked to climate change and those who do not. The primary beneficiaries are those associated with the fossil fuel industry and, more generally, the economic/political elites of most nations. These elites are dependent on market economies fueled by high levels of consumption, with this consumption being a major cause of climate change (Worldwatch Institute 2010; York et al. 2003). The advance of climate change will increase challenges to these groups, as their harmful actions become more apparent. These groups, in turn, will use their enormous power to resist such challenges. Indeed, these groups are already heavily involved in a fight to undermine meaningful action on climate change and—as suggested above—have been quite successful in this fight (Frantz and Mayer 2009; Henson 2011; Lynch et al. 2010; McCright and Dunlap 2010; York et al. 2003).

The advance of climate change will also exacerbate existing conflicts between developed and developing nations (Parks and Roberts 2008). Those in developing nations did not cause climate change, but they will suffer greatly from its effects. For example, they are far more likely to experience death, injury, and homelessness from climate-related disasters (see Global Humanitarian Forum 2009; Parks and Roberts 2008; Oxfam 2009). Consequently, developing nations will become more insistent in their demand that developed nations make large cuts in carbon emissions and provide them with aid, both to help them adapt to climate change and build sustainable economies. Developed nations, however, will be reluctant to drastically cut their emissions and provide massive aid to distant others—especially when they feel threatened by the above strains. This conflict between developed and developing nations is perhaps the major reason for the current failure to reach meaningful international agreements on climate change (Parks and Roberts 2008).

Climate change will also create new conflicts. Most such conflicts will likely center around competition over scarce resources, such as freshwater and fuel. Migration will also increase conflict, particularly when large numbers of people move into areas with scarce resources. Also, conflict will emerge as communities, states, and international organizations lose their ability to exercise effective social control. The exercise of such control prevents much conflict, since groups are encouraged to restrain themselves from acting on their immediate interests and governmental bodies mediate those disputes which do arise. In sum, the advance of climate change will undermine the ability to take meaningful action, as resources become scarce, social control weakens, and conflict increases.

The Obligation to Act

As climate change advances, the great harm that it causes will certainly foster an obligation to act. Increasing numbers of people will call for action, such action will appear increasingly desirable, and the negative emotions aroused by climate change will increase. But at the same time, this advance will provide a variety of excuses and justifications for inaction—undermining this general obligation to act. Most of these excuses and justifications were suggested above. Individuals and groups will claim that they must focus on more pressing issues, particularly those involving basic needs for food, shelter, and security, that they lack the resources to act, and/or that it is not their responsibility to act.

Further, group conflict will lead some groups to denigrate others. It is common for conflicting groups to blame one another for the problems they are experiencing and, in some cases, to dehumanize each other (Bandura 1990). This denigration will also undermine the obligation to take meaningful action, since it reduces the moral concern for those in other groups. The result, instead, will be more limited actions that benefit one's in-group and often hurt those in outgroups. The wealthy, for example, may take steps to protect themselves from the worst effects of climate change—even though doing so hurts those in other groups. For example, the wealthy may use their resources to build protected communities and secure needed resources, while others remain exposed to the worst effects of climate change. Related to this, we might also expect efforts to deny the suffering experienced by these others (see Cohen 2001).

The Interest to Act (Costs and Benefits of Acting)

Even if individuals and groups feel an obligation to act, they may not do so if the perceived costs of action outweigh the benefits. As noted above, the effects of climate change will promote a focus on immediate interests. This reduces the likelihood of a favorable cost–benefit ratio for acting on climate change, since the upfront costs of mitigation and adaptation efforts are often high, but the benefits are delayed. Further, the upfront costs will become increasingly high as climate change proceeds, while the ability to pay such costs will decline. And, as noted above, the increase in costs will be particularly high for those associated with the fossil fuel industry and for the economic/political elites in many countries—making these especially powerful groups quite reluctant to act. At the same time, the immediate costs associated with criminal or harmful responses to climate change will decline as social control breaks down.

Conclusion

The above arguments are quite bleak, suggesting that the advance of climate change will reduce rather than increase the likelihood of meaningful efforts to mitigate and adapt to it. This will occur because climate change increases strain, reduces social support and control, increases social conflict, fosters justifications and excuses for inaction, and increases the cost of meaningful action. The end result will be a Hobbesian world, with people and groups struggling to survive in a harsh environment, and the state unable to provide adequate support or exercise effective social control. At the same time, it is critical to note that there is an alternative, more hopeful narrative.

Research indicates that people are not simply motivated by self-interest; they have a prosocial side as well. Among other things, they care about others, desire to cooperate with them, and will come to their aid in times of need—even if it involves some (limited) cost (e.g., Penner et al. 2005; Sautter et al. 2011). These prosocial tendencies apply more to members of one's ingroup than to outgroups; but people have become increasingly likely over time to view diverse others as part of their ingroup (Singer 1981). Further, these prosocial tendencies are sometimes displayed even during times of hardship. For example, disaster researchers have found that individuals tend to help one another in the period immediately following disasters (Tierney 2007). Perhaps reflecting these facts, there is now much environmental concern throughout the world, even among the poor and those in developing nations (Dunlap and York 2008). So there is some reason to believe that people may be able to overcome their immediate interests and divisions, and take meaningful action on climate change. Beyond that meaningful action might also be fostered by technological advances, particularly advances that dramatically reduce the upfront costs of responding to climate change (Gore 2009).

It is at this point an open question which narrative will dominate our response to climate change. This is a topic researchers should investigate, through both case studies and quantitative research. In particular, researchers might examine how the effects described above—such as food and freshwater shortages, the loss of home and livelihood due to natural disaster, and migration in the context of scarce resources—influence the ability and motivation to take meaningful action on climate change. Such research will be limited by the fact that it cannot fully duplicate the severe effects of future climate change. Nevertheless, it can provide useful information. I suspect that evidence will be found for both narratives, with climate change stimulating both self-interested and prosocial behaviors.

The central point of this paper, however, is that it is critical to be aware of the possibility that climate change may at least sometimes have the negative effects described above. And if research finds evidence for these effects, steps can be taken to reduce or counteract them. Social and behavioral scientists have discussed ways in which we might foster environmental concern, increase the perceived and actual ability to engage in environmentally responsible behavior, create an obligation to engage in such behavior, and reduce the likelihood that people will base their actions solely on immediate self-interest (e.g., Kurz 2002; Lindenberg and Steg 2007; McCright and Dunlap 2010; Patchen 2010; Parks and Roberts 2008). And, as suggested in this chapter, criminologists have much to contribute to this discussion. In particular, we might increase the likelihood of a meaningful response to climate change by addressing certain of the criminogenic effects of such change, particularly increased strain, reduced social support and control, the development of beliefs conducive to harmful behavior, and increased social conflict (for suggestions in these areas, see Agnew 2009; Barlow and Decker 2010; Simpson and Weisburd 2010).

References

Agnew, R. (2006). *Pressured into crime: An overview of general strain theory*. Oxford: New York.

Agnew, R. (2009). *Juvenile delinquency: Causes and control*. Oxford: New York.

Agnew, R. (2011). Dire forecast: A theoretical model of the impact of climate change on crime. *Theoretical Criminology, 16*(1), 21–42. doi:10.1177/1362480611416843.

Bandura, A. (1990). Selective activation and disengagement of moral control. *Journal of Social Issues, 46*, 27–46.

Barlow, H. D., & Decker, S. H. (2010). *Criminology and public policy*. Philadelphia, PA: Temple University Press.

Brody, S. D., Zahran, S., Vedlitz, A., & Grover, H. (2008). Examining the relationship between physical vulnerability and public perceptions of global climate change in the United States. *Environment and Behavior, 40*, 72–95.

Buhaug, H., Gleditsch, N. P., & Theisen, O. M. (2008). *Implications of climate change for armed conflict*. Washington, DC: The World Bank.

Cohen, S. (2001). *States of denial: Knowing about atrocities and suffering*. Cambridge: Blackwell.

Cullen, H. (2010). *The weather of the future*. New York: Harper.

Cullen, F. T., & Agnew, R. (2011). *Criminological theory: Past to present*. Oxford: New York.

Diekmann, A., & Preisendorfer. (2003). Green and greenback: The behavioral effects of environmental attitudes in low-cost and high-cost situations. *Rationality and Society, 15*, 441–472.

Dietz, T., Dan, A., & Shwon, R. (2007). Support for climate change policy: Social psychological and social structural influences. *Rural Sociology, 72*, 185–214.

Dunlap, R. E., & York, R. (2008). The globalization of environmental concern and the limits of the postmaterialist values explanation. *Sociological Quarterly, 49*, 529–563.

Frantz, C. M., & Mayer, F. S. (2009). The emergency of climate change: Why are we failing to take action? *Analyses of Social Issues and Public Policy, 9*, 205–222.

Franzen, A., & Meyer, R. (2010). Environmental attitudes in cross-national perspective: A multilevel analysis of the ISSP 1993 and 2000. *European Sociological Review, 26*, 219–234.

Giddens, A. (2009). *The politics of climate change*. Cambridge: Polity Press.

Global Humanitarian Forum. (2009). *The anatomy of a silent crisis*. Copenhagen: Global Humanitarian Forum.

Gockeritz, S., Schultz, P. W., Rendon, T., Cialdini, R. B., Goldstein, N. J., & Griskevicius, V. (2010). Descriptive normative beliefs and conservation behavior. *European Journal of Social Psychology, 40*, 514–523.

Gore, A. (2009). *Our choice*. Emmaus, PA: Rodale.

Hayward, S. F. (2010). Public opinion about the environment: Notable shifts in recent years. http://www.environmentaltrends.org/single/article/public-opinion-about-the-environment-notable-shifts-in-recent-years.html.

Henson, R. (2011). *The rough guide to climate change*. London: Rough Guides.

Hertsgaard, M. (2011). *Hot: Living through the next fifty years on earth*. Boston: Houghton Mifflin Harcourt.

Iwata, O. (2004). Some psychological correlates of environmentally responsible behavior. *Social Behavior and Personality, 32*, 703–714.

Kolmannskog, V. O. (2008). *Future floods of refugees*. Oslo: Norwegian Refugee Council.

Kurz, T. (2002). The psychology of environmentally sustainable behavior. *Analyses of Social Issues and Public Policy, 2*, 257–278.

Lancet and University College London Institute for Global Health Commission. (2009). Managing the health effects of climate change. *Lancet, 373*, 1693–1733.

Lindenberg, S., & Steg, L. (2007). Normative, gain and hedonic goal frames guiding environmental behavior. *Journal of Social Issues, 63*, 117–137.

Lynch, M. J., Burns, R. G., & Stretesky, P. B. (2010). Global warming and state-corporate crime: The politicalization of global warming under the Bush administration. *Crime, Law, and Social Change, 54*, 213–239.

McCright, A. M., & Dunlap, R. E. (2010). Anti-reflexivity: The American Conservative Movement's success in undermining climate science and policy. *Theory, Culture, and Society, 27*, 100–133.

McKibben, B. (2010). *Eaarth: Making a life on a tough new planet*. New York: Times Books.

Oxfam. (2009). *Suffering the science: Climate change, people, and poverty*, Oxfam Briefing Paper 130. Copenhagen: Oxfam International. http://www.oxfam.org.

Page, L. A., & Howard, L. M. (2010). The impact of climate change on mental health. *Psychological Medicine, 40*, 177–180.

Parks, B. C., & Roberts, J. T. (2008). Inequality and the global climate regime: Breaking the North-South impasse. *Cambridge Review of International Affairs, 21*, 621–648.

Patchen, M. (2010). What shapes public reactions to climate change? Overview of research and policy implications. *Analyses of Social Issues and Public Policy, 10*, 47–68.

Penner, L. A., Dovidio, J. F., Piliavin, J. A., & Schroeder, D. A. (2005). Prosocial behavior. *Annual Review of Psychology, 56*, 365–392.

Raleigh, C., Jordan, L., & Salehyan, I. (2008). *Assessing the impact of climate change on migration and conflict*. Washington, DC: The World Bank.

Sautter, J. A., Czap, N. V., Kruse, C., & Lynne, G. D. (2011). Farmers' decisions regarding carbon sequestration: A metaeconomic view. *Society and Natural Resources, 24*, 133–147.

Simpson, S. S., & Weisburd, D. (2010). *The criminology of white-collar crime*. New York: Springer.

Singer, P. (1981). *The expanding circle: Ethics and sociobiology*. New York: Farrar, Straus, & Giroux.

Smith, D., & Vivekananda, J. (2007). *A climate of conflict*. London: International Alert.

Tacoli, C. (2009). Crisis or adaptation? Migration and climate change in a context of high mobility. *Environment and Urbanization, 21*, 513–525.

Takacs-Santa, A. (2007). Barriers to environmental concern. *Human Ecology Review, 14*, 26–38.

Tierney, K. J. (2007). From the margins to the mainstream? Disaster research at the crossroads. *Annual Review of Sociology, 33*, 503–525.

Worldwatch Institute. (2010). *2010 State of the world: Transforming cultures*. New York: W. W. Norton.

York, R., Rosa, E. A., & Dietz, T. (2003). Footprints on the earth: The environmental consequences of modernity. *American Sociological Review, 68*, 279–300.

Chapter 3
Heading Toward a New Criminogenic Climate: Climate Change, Political Economy and Environmental Security*

Pete Fussey and Nigel South

Introduction

Over the past 20 years, much scientific enquiry has focused on the origins and drivers of climate change, whilst economic debate has informed discussion around pre-emptive and adaptive responses to what is set to become the defining issue of the twenty-first century. Perhaps surprisingly, despite the intensity of climate-related debates, there remains much that is certain. For example, very few question the existence of climate change and, also, its anthropogenic causes; of those that do, even fewer are seen as credible. The debate instead rests on the questions of "how much" and "how fast"? On the other hand, amid rapid ecological and technological changes there is also considerable uncertainty about what the future will look like and, as an important corollary, the optimal trajectories towards a sustainable future.

The Kyoto protocol, part of the UN Framework Convention on Climate Change, adopted in 1997 and in force from 2005, provides some certainty of assurance that the problem is recognised and that states can move toward some agreement about the need to reduce carbon-based greenhouse gases (GHG) to 5% lower than 1990 levels by 2050. However, signing up to Kyoto is optional, there are no regulatory teeth to push nations to opt-in, and signing does not necessarily mean ratifying and committing to the measures necessary to cap emissions. Gardiner (2011) for example, would see Kyoto less as an outstanding beacon of what can be achieved and more as something of a smart licensing arrangement enabling wealthy, developed nations to continue to support existing systems and lifestyles with little or no impact on emissions or progress on related fronts: business as usual. Nonetheless,

*Some of the discussion contained within this chapter is informed by research for the ESRC and EPSRC funded project EP/I0005943/1 "Resilient Futures".

P. Fussey • N. South (✉)
Department of Sociology, University of Essex, Colchester CO4 3SQ, UK
e-mail: soutn@essex.ac.uk

R. White (ed.), Climate Change from a Criminological Perspective,
DOI 10.1007/978-1-4614-3640-9_3

one clear ancillary benefit of the Kyoto agreement has been to focus academic research toward a common temporal framework, imposing some order on an unwieldy and growing academic body of work. In doing so, cross-disciplinary analyses become easier to engineer (and constitute an endeavour that criminology is belatedly engaging in: South (1998, p. 226); Farrall et al. 2012). The fact that the Kyoto goals will either be met or be missed has, in turn, led to various polarised visions of the future, often situated on one or other side of the utopian–dystopian divide (examples abound in numerous scenario exercises: see Shell 2009; Chap. 2).

One inevitable feature of this future is the restricted availability of the fossil fuels that currently underpin the carbon economies of the developed world. Moreover, there is broad consensus that the deleterious effects of climate change will be distributed unevenly—an issue that speaks to the heart of sociological analysis yet, curiously, currently remains at the margins of the discipline. Here, the coming "climate divide" will represent a further extension of the inequitable state of the affairs of humanity and the planet, one in which the conditions producing climate change are contributed to most overwhelmingly by rich consumer societies but which will impose the greatest costs and resultant miseries on the already poor and newly developing nations.

This chapter draws on this mix of certainties and uncertainties as a framework for consideration of the criminogenic features of climate change. It commences with a brief discussion of key drivers and outcomes of climate change before suggesting a number of social and criminological implications, with particular relation to the political economy, environmental security and eco-social change. The chapter concludes by drawing these themes together and, building on observations from Perelman (2008) and the body of work attending to "resilience", arguing that sustainability is a crucial component of future security. The chapter also provides an initial response from criminology to Urry's (2011) recent and urgent call to instate "social", and particularly sociologically informed, analyses within mainstream discourses around climate change, to place it alongside the thus far dominant disciplinary behemoths of science and economics. The inclusion of the "social" realm is critical and crucial: it filters, catalyses and mediates the key drivers of climate change and absorbs its impacts. Criminology is already a meeting point for various perspectives, methods and theoretical approaches and can bring to the study of climate change a rich tradition of understanding the implications of social, cultural, economic and political forces.

Future Proof? Drivers of Climate Change and the Context for Criminality

The causes and impacts of climate change are inherently complex and, in some respects, contested. Clearly, a detailed exegesis of causes and effects lies outside the scope of this (or any) single chapter. Here, broader debates surrounding the drivers and impacts of climate change are reduced to a number of base elements. These are selected both on the principle of "certainties" and "uncertainties" as outlined above, and for their potential sociological and criminological resonance. These are population growth, urbanisation, mobility and fossil fuel exhaustion. Together, these represent a series of interlocking trends that will radically impact upon the planet and the social lives of its inhabitants and, in turn generate a range of criminogenic corollaries. These broad changes are now discussed in turn.

The global population is growing. According to the UN's influential projection of demographic change towards 2300, today's world population of 6.9 billion is (almost certainly conservatively) estimated to rise to just less than nine billion by 2050 (UN ESA 2004).[1] Key here are the multiple and uneven drivers of this growth. In the global North, increased life expectancy will outpace the rate of total global population growth (UN ESA 2005). Yet it is the developing regions of the world that will absorb the vast majority of the planet's additional inhabitants (UN ESA 2006). The environmental impacts of such changes are considerable. Most obvious is the simple equation that the existence of more people will result in further consumption of carbon-based resources. The population will increase in places of scarcity thus amplifying inequalities (at local and global levels), intensify competition for already scarce energy resources and exert considerable pressures on existing overstretched infrastructures.

At the same time, United Nations Habitat (2009) predict that this demographic growth will be accompanied by an accelerated process of urbanisation, stimulating a near doubling of the world's urban population by 2050. Similar to general patterns of demographic expansion, urban growth will be concentrated in the global South, and will largely occur outside of formal planning regimes and with few controls on the quality and security of tenure. These pressures are likely to be particularly acute in Africa, where urban populations are set to triple during this period (United Nations Habitat 2010). In environmental terms, urbanisation is particularly harmful. Currently, cities accommodate just over half the global population (United Nations Habitat 2009) yet consume three quarters of the world's energy and discharge a similar proportion of the planet's GHGs (Urry 2011). Moreover, the amplified inequalities, enhanced competition for resources, rapid demographic change and accelerated urbanisation brought by these changes are not only resource-intensive processes but also highly criminogenic (South 2010).

A corollary of population growth is increased mobility. Contemporary lifestyles in the developed North are often predicated on extensive "network capital", involving extensive personnel and business connections across dispersed international settings, and requiring high levels of geographical mobility (Urry 2011). Such distanciated relationships increasingly strain transport infrastructures, are carbon intensive and constitute a rapidly growing trend. Whilst road transportation

[1] Less agreement exists over the trajectories of global population after 2050. Some predict a rapid halt to population growth followed by decline (Jackson 2004). Others project alternative pathways of growth, stabilisation and a degree of decline (UN ESA 2006). Such uncertainties and likely change further complicate long-term planning exercises.

contributes to 74% of global GHG emissions (a rate that has grown by 27% since 1990) (Arup 2009), it is aviation that contributes the most egregious effects. Not only does air travel constitute the most CO_2 per kilometre of any mode of mainstream travel, but it also taints the earth's upper atmosphere with nitrogen-based pollutants that have an exaggerated climactic impact at that altitude. Aviation-related emissions have sharply increased by 44% since 1990 (Arup 2009), forming a trend that is set to continue (and is currently partly driven by the exponential growth of the industry in India, particularly within the short-haul budget airline sector). Yet, despite the continuance of this trend, there is a strong possibility that, at some stage, a crisis point will occur. There is apparently little that can be done to alter the environmental harm of aviation. Aircraft can be made lighter and fuels diluted but commercial flight will still rely on fossil fuels to power inefficient airborne trajectories of continuous thrust. Despite the publicity surrounding the development of new rail infrastructure, such as the UK's "High Speed 2" (HS2) line approved during January 2012, internal air travel arguably involves less new infrastructure (HS2 will not be completed until 2026), is more flexibly deployed, carries significantly more passengers and, globally, is growing at a much faster rate. Whilst the stigmatisation of frequent fliers by environmentalists may do a little to reverse this trend—and the power of critical questioning of "everyday" environmentally damaging behaviours should not be overlooked (Agnew 2013; Babcock 2009, pp. 13–16)—future scarcity of fossil fuels and attendant inflation of air travel costs will likely have greater impact. Reduced frequency of movement and new geographical arrangements based on proximity are likely to have extensive sociological implications (see Adey 2010; Elliot and Urry 2010 for excellent analyses of such implications). Countervailing trends of abstract virtual co-presence and new forms of localism here are set to affect fundamental approaches to social interaction. Wholesale movements towards cloud computing[2] may simultaneously place information and our virtual selves, to paraphrase Foucault (1980), everywhere and nowhere. In criminological terms, intensified reliance on virtual interconnectivity increases the vulnerability of our various networks, communications and interdependencies to cybercrime and online fraud. Concerning the related outcomes of reduced mobilities and enhanced community level efficacy, the operation of informal social controls, community divisions and widening inequalities (in relation to access to transport, for example) are all staples of criminological enquiry. For some, of course, the operation of illegal markets will ensure that mobility will not be restricted and access to fuel and transport will be as available to those in positions of privilege and power as other resources tend to be in times of scarcity or prohibition (Hayman and Brack 2002, p. 7).

[2]Cloud computing involves the relocation and storage of ICT software and data to the virtual realm. Increasingly, information and software applications are stored online and synchronised to multiple machines rather than physically and separately stored on each individual device. *Dropbox* and Apple's *iCloud* and *iTunes Match* services are popular consumer-focused examples of cloud computing.

These changes will occur during a period in which the resources that have been central to the development of high-carbon economies will be significantly constrained. The future scarcity of fossil fuels, exemplified by debates around whether peak oil production has passed (and thus we are to experience exponential reductions in available oil supplies), will undoubtedly give rise to significant social, as well as environmental, implications. Given that the impacts will be felt internationally, nationally and locally, affecting individual lifestyles and institutional practises, it is difficult to overstate the importance of fossil fuel resource constraints in the coming decades.

At the time of writing (2011), there are several ways this issue may be addressed, all of which have varying social and criminological implications. First, and most likely, oil extraction will continue, but the yields will be smaller and they will be accessed via methods that pose greater risks to the environment. One such strategy is deep-water oil "exploration", largely concentrated in the coastal mid-Atlantic and Gulf of Mexico, and aimed at extracting oil at more than 400 m below sea level. The environmental catastrophe following BP's use of this technique and (by their own admission) failure to effectively manage the attendant risks during the 2010 *Deepwater Horizon* episode demonstrates the considerable dangers of this approach and the harms that can follow. Continued reliance on fossil fuels at current levels also entails significant criminological implications. These exist at different levels of action. As the above example demonstrates, high-risk extraction techniques will inevitably result in repeated environmental costs. Embedded within this dynamic are key criminological concerns of "harm" and of corporate crime. Indeed, it is not difficult to assert that such oceanic desecration contravenes the sprit of criminal law, even if the letter remains woefully inadequate in such matters. For example, it is also worth noting here the criticism levelled at BP and its partners by the US Oil Spill Commission established by Barak Obama to investigate the disaster. This centred on the shortcuts to save time and money: *Decision making processes ... did not adequately ensure that personnel fully considered the risks created by time- and money-saving decisions* (OSC 2011, p. 125 emphasis in original). Saving time and money are, of course, key to the process of capital accumulation and constitute the repeated causes of corporate killing and other serious transgressions (Ruggiero and South 2010).

Other potential responses to fossil fuel depletion and GHG harms exist, however. Typically, proposals fall into two overlapping categories. The first concerns climate change adaptation/mitigation and involves schemes to prevent the onset of global warming, or to reduce its deleterious effects. Examples include sequestration approaches such as Carbon Capture and Storage schemes to remove carbon dioxide from the atmosphere, or large-scale geo-engineering initiatives (such as oceanic fertilisation).[3] The second involves more familiar supply and demand narratives respectively focusing on renewable energy and on reducing the demand for energy

[3] Sequestration approaches generally involve the removal of GHGs—either at their source of emission or later extracted from the atmosphere—which are then stored, sequestered, indefinitely. Oceanic fertilisation comprises a range of alternative strategies such as the "fertilisation" of small organisms, such as plankton, which absorb carbon-based atmospheric pollutants.

(as seen in the extensive public campaigns in post-Fukushima Japan). Leaving aside the first approach, which largely relies on as yet undeveloped technologies, the social implications of the latter approaches are significant. In the first instance, as the UK Government's Department for Energy and Climate Change's (DECC) recent "2050 Pathway Analysis" initiative relates, reaching the 80% GHG emission reduction targets via renewable energy and demand reduction requires significant commitment to change across a number of domains. In particular, they require a step change in current approaches to transport, manufacturing, energy consumption and energy generation. Notable future social impacts include the fallout from geo-political tensions surrounding the importation of energy or, alternatively, a re-weighting of governance arrangements towards the local as micro-generation of power becomes more widespread.

Harms and Impacts: Political Economy, Environmental Security and Eco-Social Change

Whilst these specific trends and initiatives entail a variety of sociological implications, there are unquestionable macro-level impacts for the overall process of climate change. In the first instance, as already noted, existing inequalities at local, national and global levels are set to intensify. Not only will those with the least be most susceptible to climate change-induced harms, but they are also likely to be the least equipped to deal with such catastrophes. As reported in the *New York Times* (Revkin 2007) Rajendra K. Pachauri, chairman of the United Nations Climate Panel observed that "The inequity of this whole situation is really enormous if you look at who's responsible and who's suffering as a result". While life expectancies increase in the global North, less developed countries may suffer even greater health problems as insects and other agents carrying vector-borne diseases thrive in conditions of rising temperature and humidity. The impact of climate change on future gross health inequalities has been recognised by public health specialists and Costello et al. (2009, p. 1604, col 1) note that this raises the issue of intergenerational justice: "The inequity of climate change—with the rich causing most of the problem and the poor initially suffering most of the consequences—will prove to be a source of historical shame to our generation if nothing is done to address it".

Climate change also produces cruel ironies. We live in a world where water is (almost) everywhere, yet (alongside air) it is the most precious commodity on the planet. The consequences of not having access to it are poverty, starvation, disease and death while in other locations water can be unpredictably over-abundant leading to risks of flooding and pollution of clean water—again with consequences that include disease and death (Hemson et al. 2008). Rising sea levels, as in the case of encroachment over the Tuvalu islands and the inundation of low-lying areas such as the Ganges Delta, as well as the onset of extreme weather events across the globe are contrasted by the desertification of entire landscapes and restricted access to drinking water elsewhere. Regarding the latter, as global temperatures rise and water is drawn from the land, "mega" droughts, and their impact on food production and the availability of drinking water, have been mooted as one of the key future concerns for human societies (see Arup 2009).

In the future, not only will droughts become more likely, frequent and severe, but also paleoclimactic evidence demonstrates that, throughout the earth's history, there is a propensity for droughts to last for decades (Arup 2009). This should stimulate policy and practise concerned with effective water use and conservation. At present, in developed nations with well established water supplies, water is taken for granted and misused in various ways as societies indulge in behaviours that increase pollution and require wasteful over-use in industry and the home. Water has also been exploited as a profitable commodity. This occurs most obviously through the need to pay public or private bodies for collecting, purifying and supplying water but also—in recent decades—through the phenomenal growth in the market for bottled water. As Wilk (2006) observes, in the past 30 or so years, water has been pulled between being a free public good and a privatised commodity with the latter development producing popular protests in countries as diverse as Bolivia and Canada. As fresh water becomes scarcer, such profits are set to increase, as are inequalities in the ability to access basic essential resources.

Global markets already play a role in driving and exacerbating the impacts of climate change in ways that lead to contestation over resources and while this contestation is not new, the commodification of nature (the ability to produce profit from both the otherwise freely available as well as from the increasingly scarce) has become crucial to the way in which capitalism and world trade can function in a "business as usual" way. As White (2003, p. 150) observes:

> global capitalism … reaches into the very essence of the land, the air, the water and leaves no place and no person untouched by its pervasive influence. Yet, even in the midst of acknowledged ecological crisis, many writers turn to population or to technology as the source of the problem, and its solution. The tendency is to assume "business as usual" and to tweak the system around the edges

Interestingly, the choice between "business as usual" and the opportunity to fund an alternative approach has been presented in stark and simple form by the discovery of a new oil field that may contain around one billion barrels beneath one of the last truly undisturbed areas of wilderness in the world. This is the Yasuni national park in Ecuador, which is thought to have more species of plants, animals and insects per hectare than any other place on the planet (Vidal 2011, p. 18). The Ecuador government and the UN Development Programme are supporting a proposition put forward by the former Oil Minister Professor Alberto Acosta that—in the context of global acknowledgement of "ecological crisis"—essentially offered a "deal", proposing on the one hand, "a revolutionary scheme to leave the oil in the ground in perpetuity in return for half of its value from the rich countries of the world", or, the alternative of familiar recourse to environmental desecration via the "business as usual" model (Vidal 2011).

This proposition is quite explicitly one that recognises the need to de-couple economies from current carbon dependencies and instead to value the diversity and future of the planet. Unfortunately, neither "politics as usual" nor the avowedly "green" self-regulation measures of energy industries show signs of providing strong support in this direction (South 2011). At the heart of this is the global interplay of harmful commercial competitiveness and the nationalist priorities of security policies which, when concerned at all about the environment, are concerned with protecting lines of access for home consumption rather than the sustainability and availability of resources for the future.

Complicity in the harms following the "business as usual" model is not confined to the realm of the private sector. The directions taken by the United States as (at present) the major global consumer economy have been significant in the arenas of state-corporate response as well as environmental security policy and action. For example, Lynch et al. (2010) apply the concept of state-corporate crime to the actions of the G.W. Bush administration in rejecting the need for anti-global warming strategies and support for "business-as-usual". Here, state-corporate crime is not perpetrated with the specific intention of doing harm but prevention of harm is not the over-riding primary goal either. Rather, citing Friedrichs (2002), they point out that "because such acts advance the combined interest of state-corporate actors in the accumulation and maintenance of power and wealth, state-corporate crime will impose the risk of harm to citizens, consumers, and/or workers". Precisely because of the powerful interests involved, the activities in question may be seen by critics as harmful but cannot be legally or technically defined as criminal. Once again, such contravention of the spirit, yet not the letter, of the law demonstrates the urgency of reconsidering extant judicial notions of harm. This "openly environmentally hostile political context" gave rise to occasions of irresponsible disregard for problems of pollution caused by relaxation of controls over the operation and building of coal fired power plants and, also, ultimately to outright violations of existing national polices and federal laws (until changes to these became operational). As Lynch et al. (2010, p. 227) summarise, "Because of the immense power the US wields internationally, the Bush Administration was able to forestall implementation of international treaties on global warming and produce domestic policies that contributed to rather than impeded the progress of global warming" and did so in close collaboration with corporations that engaged in dis-information, "information laundering" and misrepresentation of scientific facts. While the effects of climate change can be shown to be criminogenic it seems to also be the case that denial of the facts of climate change can lead, as the analysis of this collusion points out, to "deceit, corruption, deviance and even crime" (Lynch et al. 2010).

The purpose of all this is to undermine criticism of and maintain a "business-as-usual" mindset—the normality of carbon-profligate lifestyles. In the political discourse of the Bush Administration and corporate collaborators this would be about defending economic interests and "the preservation of a way of life". In this sense, maxims of "national security" and "the national interest" might be mobilised but where this becomes more complex is that in the case of finite resources the right and the capacity to consume in increasing amounts "now" does nothing to secure the ability to do so "in the future". Such notions of security thus entirely ignore issues of intergenerational equity and how present ambitions for security and well-being will be denied to future generations. In other words, such security is not sustainable and indeed the policies and practises supporting it could be construed as being criminally irresponsible toward future generations. At the same time, such obfuscation, denial and inactive witnessing of environmental crises connect with social science analyses of the deleterious impact of the bystander and of action in "bad faith" (see Cohen 2001). Hence a model of governance and security that is environmentally sensitive is needed and discussions around the concept of environmental security have made some significant contributions here (South 2012).

Understanding the Criminological Relevance of These Social Changes

Abbott (2008) has written of the "uncertain future" that climate change holds for the world in general and for policies and systems related to law enforcement and national security. Broadly, the kind of eco-social forces or socio-economic impacts of climate change that we have been describing, such as loss of infrastructure, resources scarcity, and mass displacement of peoples (Abbott 2008, pp. 6–7) have various criminogenic influences and hence have implications for law enforcement and policing: demands for greater border security, policing of new legislation, and responding to natural disasters but of particular interest here, changes in rates and types of crime. Here, Abbott (2008) links crime to climate-induced migration of peoples from areas suffering the worst hardships caused by climate change.

In respect to destination countries, migration and new settlers can bring many familiar features of dislocated and displaced communities and individuals—they may travel with or rely on savings that are transmitted by illegal, subterranean banking methods, and these may be invested in objects that are more resistant to loss of value than some other items which may involve smuggling, avoiding taxation, investment in drugs or other contraband. New settler groups that are excluded from the mainstream society have often seen some members adapt through legitimate means but others create organised crime opportunity systems. Belief systems, religions, cultures and matters of honour can underpin violence between groups in the community and within the domestic sphere and lead to prejudice, victimisation and hate crimes committed by both the host and incoming communities. Such processes imply a climate-related interpretation of Taylor's (1999) analysis of "drawbridge mentalities"—the assertion of (arbitrary) insider/outsider binaries and exclusionary policies—and their consequences at a time of economic downturn. Moreover, Bauman's (1998) characterisation of globalisation-driven mobilities, cleaving urban populations into dispossessed, over-policed and immobile vagabonds on the one hand and mobile, protected and affluent tourists on the other, may resonate in the form of increased social control visited upon the impoverished victims of climate change.

Such tensions are not only confined to the destination countries. Given the aforementioned particularly acute impacts of climate change on the global South, these conditions not only provide powerful drivers of migration, but also generate significant criminogenic forces. Here, for example, the impact of drought and the interconnected effects of climate change, everyday human action and political tensions and antipathies have been analysed in relation to displacement and genocide in Darfur:

> By the mid-1980s, the intertwined processes of desertification and famine aggravated disputes over land and water and intensified the socially constructed, racially tinged division between Arabs and other Africans. The causes of the conflict in Darfur are clearly a mixture of environmental and political forces. Understanding the interconnection of these forces and how they are played out is central to explaining state-led genocide through elimination of non-Arab groups in this African setting. ...Desertification is an environmental hazard and challenge caused both by natural climate change and overgrazing and farming (Hagan and Kaiser 2011, p. 5)

Drought induced food scarcity has devastating consequences in terms of human suffering but can also stimulate illicit markets in food and medicines with aid diverted and misappropriated. At the other extreme, flooding destroys food crops, pollutes and spreads disease and similarly gives rise to the need for aid. Again it can also lead to exploitation of those seeking to escape such conditions, in the shape of human trafficking, predatory banditry or extortion of "protection money". Regarding the latter for example, Walters (2011) discusses how food aid can be used as a lever by powerful corporations and hegemonic states to impose Genetically Modified Organism (GMO) technology on countries that are unwilling and unsuited to its arrival. In a broader context, Smith and Vivekananda (2007) have identified "double-headed" threats of both climate change itself and of corollary violent conflicts. Here, they identify 46 countries, accommodating 2.6 billion people, deemed at a high risk of experiencing climate-change related armed conflict.

At a more general level, other forms of crime may increase or see some change in pattern or frequency following more extreme or changeable weather conditions. For example, domestic burglaries and other opportunistic thefts are weather related due to tendencies to leave windows and doors open or unsecured in hotter weather. Additionally, crimes of violence have been shown to rise during heatwaves (albeit for a variety of reasons including increased alcohol consumption in hot weather, Hughes et al. 2004). The natural world provides the basics for life but it also produces the luxuries that attract premium values—particular kinds of foods, drinks, materials for fashionable clothes and accessories. In the event of species decline or even extinction, such goods will become even more rare and hence more expensive, and as they are likely to be judged as worthy of conservation then illegal trades will open up to circumvent such controls. The impact of hurricanes and floods (such as Katrina and the Mississippi) affecting mainland United States have signalled the vulnerability of even wealthy and technologically advanced nations to the power of nature. In such circumstances systems and order break down and "ordinary people" will engage in looting of stores for food and water that is not being provided by other means. At both home and abroad, those with an interest in the security of the state have taken note. Hence as a New York Times story (Broder 2009) noted:

> The changing global climate will pose profound strategic challenges to the United States in coming decades, raising the prospect of military intervention to deal with the effects of violent storms, drought, mass migration and pandemics, military and intelligence analysts say. Such climate-induced crises could topple governments, feed terrorist movements or destabilise entire regions, say the analysts, experts at the Pentagon and intelligence agencies who for the first time are taking a serious look at the national security implications of climate change.

According to Abbott (2008, pp. 7–8) the "security consequences" of climate change that might stand out are civil unrest, inter-communal violence and international instability. Whether in cases of looting following local disaster or these grander instances of social and political breakdown, climate change will have significant though as yet relatively under-explored impacts.

Conclusions: Meaning and Mitigation

This chapter argues that climate change is intrinsically connected to criminological concerns. This relationship can be observed in relation to the *drivers* of climate change and the criminality associated with the commodification of nature and, also, in relation to the *impacts* of climate change. At the same time, the escalation of global warming will generate fundamental questions over the application and, crucially, direction of social and other control strategies. In particular, extant orthodoxies of control have proved deficient at regulating the economic and political forces that have driven climate change and other environmental harms. Yet, mechanisms of social control have proved incredibly adept at penalising those on the margins of society and those undertaking transgressive behaviour as a strategy of economic survival. As the significant social upheavals ushered in by climate change develop, the maintenance of such orthodoxies are set to amplify its deleterious effects on those who are already the most heavily disadvantaged. For these reasons, it hardly needs stating that a step change in the approach to regulating and controlling the drivers of climate change and limiting its impacts upon vulnerable populations is needed. What is often overlooked, however, is the importance of decoupling our responses from attempts to maintain "business as usual" approaches, given their catastrophic failure in this regard. As such, market-based schemes, such as carbon trading, risk continuing and reproducing structural deficiencies that underpin much of the current crises.

At the same time, this chapter and other research has highlighted how carbon-profligacy is not solely located within the realms of corporate irresponsibility but is an embedded feature of late-modern living in the developed world. One key to addressing such broader interrelationships between criminology and climate change is an understanding of complexity, appreciation of cascading effects, and cognizance of what can be anticipated and what cannot. Such concerns have recently been articulated in the OECD's (2011, p. 3) interpretation of the "global shocks" model which, "takes account [of] cascading risks that become active threats as they spread across global systems, whether these arise in health, climate, social or financial systems". Thus, climate change imposes myriad impacts that in turn generate additional crises of both anticipated and unanticipated natures. For example, extreme weather events have multiple impacts at macroeconomic (such as Japanese currency markets, following the 2011 Tsunami, see Yokoyama 2011), community and individual levels (such as intercommunity tensions and psychological impacts after Hurricane Katrina in 2005, see Brunsma et al. 2010). Although OECD models are (unsurprisingly) rooted in economic and business-as-usual approaches, and scaled towards the transnational level of analysis there is much that can be applied to criminological enquiry, particularly in relation to the dynamic and interlocking nature of climate change-induced crises. As such, mitigation responses may address different features of this complexity simultaneously and, also, operate at different levels of action.

At a macro-level, as climate change becomes an ever more central part of national security calculations and international relations negotiations, these ways of thinking and operating must be more proactively green and sustainability-oriented otherwise they will prove to be self-defeating. Neither national well-being nor international cooperation can be secure and thrive in the uncertain future that climate change will shape unless, as Abbott (2008, p. 11) suggests, the accompanying opportunities for prevention, mitigation and adaptation are vigorously pursued. At the same time, conceptual and policy developments could explore ways to unite the "green" of sustainability with the "blue" of security (Perelman 2008) within broader discourses of resilience. Since 9/11 expenditure on countering terrorism and insurgencies has outstripped, and in many respects replaced, attention to mitigating natural and climate-change induced hazards. There are clear benefits to this approach, not least in respect of finding common ground and shared political will between often divergent and sometimes dogmatic security and sustainability agendas. However, it is also important to stress that, given the enormous social upheaval and criminogenic corollaries brought by climate change, sustainability and security cannot meaningfully be seen as entirely separate: sustainability *is* security and vice versa.

References

Abbott, C. (2008). *An uncertain future: Law enforcement, national security and climate change*. Oxford: Oxford Research Group.

Adey, P. (2010). *Aerial life: Spaces, mobilities, affects*. Chichester: Wiley-Blackwell.

Agnew, R. (2013). The ordinary acts that contribute to ecocide: A criminological analysis. In A. Brisman & N. South (Eds.), *International handbook of green criminology*. London: Routledge.

Arup. (2009). *Drivers of change*. London: Prestel.

Babcock, H. (2009). Global climate change: A civic republican moment for achieving broader changes in environmental behavior. *Pace Environmental Review, 26*(1), 1–26.

Bauman, Z. (1998). *Globalization: The human consequences*. Cambridge: Polity.

Broder, J. (2009, Aug 9). Climate change seen as threat to U.S. security. *New York Times*.

Brunsma, D., Overfelt, D., & Picou, S. (Eds.). (2010). *The sociology of Katrina: Perspectives on a modern catastrophe*. Plymouth: Rowman & Littlefield.

Cohen, S. (2001). *States of denial, knowing about atrocities and suffering*. Oxford: Polity.

Costello, A., Abbas, M., Allen, A., Ball, S., Bell, S., Bellamy, R., Friel, S., Groce, N., Johnson, A., Kett, M., Lee, M., Levy, C., Maslin, M., McCoy, D., McGuire, B., Montgomery, H., Napier, D., Pagel, C., Patel, J., Antonio, J., de Oliveira, P., Redclift, N., Rees, H., Rogger, D., Scott, J., Stephenson, J., Twigg, J., Wolff, J., & Patterson, C. (2009). Managing the health effects of climate change. *Lancet, 373*, 1693–1733.

Elliot, A., & Urry, J. (Eds.). (2010). *Mobile lives*. London: Routledge.

Farrall, S., French, D., & Ahmed, T. (Eds.). (2012). *Climate change: Exploring the legal and criminological consequences*. Oxford: Hart.

Foucault, M. (1980). In C. Gordon (Ed.), *Power/knowledge: Selected interviews and other writings 1972–1977*. Brighton: Harvester.

Friedrichs, D. (2002). State-corporate crime in a globalized world: Myth or major challenge? In G. W. Potter (Ed.), *Controversies in white collar crime*. Cincinnati: Anderson.

Gardiner, S. (2011). *A perfect moral storm: The ethical tragedy of climate change*. Oxford: Oxford University Press.

Hagan, J., & Kaiser, J. (2011). The displaced and dispossessed of darfur: Explaining the sources of a continuing state-led genocide. *British Journal of Sociology, 62*(1), 1–25.

Hayman, G., & Brack, D. (2002). *International environmental crime: The nature and control of environmental black markets*. London: Chatham House.

Hemson, D., Kulindwa, K., Lein, H., & Mascarenhas, A. (Eds.). (2008). *Poverty and water: Explorations of the reciprocal relationship*. London: Zed Books.

Hughes, S., Bellis, M., Bird, W., & Ashton, J. (2004). *Weather forecasting as a public health tool*. Liverpool: Centre for Public Health, Liverpool John Moores University.

Jackson, N. (2004). When the population clock stops ticking. In R. White (Ed.), *Controversies in environmental sociology*. Cambridge: Cambridge University Press.

Lynch, M., Burns, R., & Stretesky, P. (2010). Global warming and state-corporate crime: The politicization of global warming under the Bush administration. *Crime, Law and Social Change, 54*, 213–239.

OECD. (2011). *Future global shocks: Improving risk governance*. Paris: OECD Publishing.

Oil Spill Commission (OSC). (2011). *Deep water: The Gulf Oil Disaster and the Future of Offshore Drilling Report to the President*. Washington, DC: National Commission on the BP Deepwater Horizon Oil Spill and Offshore Drilling.

Perelman. (2008). *Infrastructure risk and renewal: The clash of green and blue*. Fairfax, VA: Public Entity Risk Institute.

Revkin, A. (2007, Apr 1). Poor nations to bear brunt as world warms. *New York Times*.

Ruggiero, V., & South, N. (eds) (2010). Critical criminology and crimes against the environment (and essays by others). *Green Criminology* (spl issue), *Critical Criminology, 18*, 4.

Shell. (2009). *Shell global scenarios to 2050*. The Hague: Shell International BV.

Smith, D., & Vivekananda, J. (2007). *A climate of conflict: The links between climate change, peace and war*. London: International Alert.

South, N. (1998). 'A green field for criminology?: a proposal for a perspective' in *For a Green Criminology*, special issue of *Theoretical Criminology, vol. 2, 2*. London and Thousand Oaks, Ca., Sage. pp. 211–234

South, N. (2010). The ecocidal tendencies of late modernity: Transnational crime, social exclusion, victims and rights. In R. White (Ed.), *Global environmental harm: Criminological perspectives*. Willan: Cullompton.

South, N. (2011). Environmental offending, regulation and the legislative balancing act. In J. Gobert & A.-M. Pascal (Eds.), *European developments in corporate criminal liability*. London: Routledge.

South, N. (2012). Climate change, environmental (in)security, conflict and crime. In S. Farrall, D. French, & T. Ahmed (Eds.), *Climate change: Legal and criminological implications*. Oxford: Hart.
Taylor, I. (1999). *Crime in context*. Cambridge: Polity.
United Nations Department of Economic and Social Affairs (UN ESA). (2004). *World Population to 2300*. New York: United Nations.
United Nations Department of Economic and Social Affairs (UN ESA). (2006). *World population prospects: The 2006 revision*. New York: United Nations.
United Nations Habitat. (2009). *Global report on human settlements 2009: Planning sustainable cities*. Nairobi: UN Human Settlements Programme.
United Nations Habitat. (2010). *The State of African Cities 2010: Governance, inequalities and urban land markets*. Nairobi: UN Human Settlements Programme.
Urry, J. (2011). *Climate change and society*. Cambridge: Polity.
Vidal. (2011, Aug 14). Ecuador asks: how much is the rainforest worth? *The Observer, New Review*, pp. 18–19.
Walters, R. (2011). *Eco crime and genetically modified food*. London: Routledge.
White, R. (2003). Environmental issues and the criminological imagination. *Theoretical Criminology, 7*(4), 483–506.
Wilk, R. (2006). Bottled water: the pure commodity in the age of branding. *Journal of Consumer Culture, 6*, 303–25.
Yokoyama, M. (2011, Aug 9). *East Japan major earthquake and tsunami, and nuclear power plant accidents as man-made disaster*. Paper presented at the International Society of Criminology World Congress, Kobe, Japan.

Chapter 4
The Cultural Silence of Climate Change Contrarianism

Avi Brisman

Introduction

Climate change has been called "[t]he overriding environmental issue of these times" (Editorial 2008) and "the most pressing and important international issue facing humanity today" (White 2009, p. 11). Al Gore (2006, p. 8) has asserted that "[n]ot only does human-caused global warming exist, but it is also growing more and more dangerous, and at a pace that has now made it a planetary emergency"; President Barak Obama announced at the Governors' Climate Summit in 2008 that "[f]ew challenges facing America, and the world, are more urgent than combating climate change" (quoted in Eilperin 2008).

Increasingly, criminologists of different theoretical orientations have begun to examine the present or "already experienced" effects of climate change, as well as possible future impacts—endeavors which have, no doubt, contributed to the inspiration for this book. For example, White (2009) has considered the potential social conflicts surrounding climate change, including conflicts over environmental resources (e.g., water, food, fish); climate-inducted migration of peoples; loss of territory and border disputes (caused by receding coastlines and desertification); conflicts over differential exploitation of resources (e.g., indigenous people and biopiracy, subsistence versus industrial production, conflicts over energy supply); and conflicts over transference of harm (e.g., cross-border pollution, transborder movement of toxic waste, circulation of pollution and waste). White (2009) has also outlined the criminalization and regulation of activities relating to carbon emissions—specifically, the criminalization of carbon gas emission and the regulation of carbon emission trading.

A. Brisman (✉)
Department of Criminal Justice, 467 Stratton, Eastern Kentucky University, 521 Lancaster Ave., Richmond, KY 40475, USA
e-mail: abrisma@emory.edu

R. White (ed.), Climate Change from a Criminological Perspective,
DOI 10.1007/978-1-4614-3640-9_4, © Springer Science+Business Media New York 2012

Lynch and Stretesky (2010) have charted the expansion of environmental harms due to climate change, as well as the unequal distribution of climate change harms, and have contemplated climate change as state-corporate crime, the intersections of climate change and criminal justice policy, and new forms of crime or black markets that develop around items people desire or need in a world impacted by climate change. In a separate publication, Lynch et al. (2010) have examined the cooperative political and corporate strategy that characterized climate change policy under the George W. Bush administration, and have proposed that this extensive collusion between the Bush Administration and the corporate sector constitutes a form of state-corporate crime. More specifically, they have argued that the Bush Administration's failure not only to address climate change but to even recognize its importance as a social problem with far-ranging implications is a "crime against the human and non-human habitants of the world" (Lynch et al. 2010, p. 215).

Emphasizing climate change as a critical criminological concern, Ruggiero and South (2010) have highlighted how climate change produces dramatically unequal impacts across geographic regions, inflicting greater damage on developing countries and preindustrial societies, and presenting the greatest threats to the poor and the unborn. Ruggiero and South (2010, p. 246) also present climate change as part of the larger issue of limiting harm to the environment, and urge critical criminologists "to take up the idea of the defense of the earth alongside advocacy of human rights as a global call for the future."

Wachholz (2007), drawing on Denton (2000, 2002), explores the gender-differentiated impacts of climate change—many of which are embedded in the deep roots of gender inequality—such as women's experiences within various forms of paid and unpaid labor (e.g., labor related to subsistence farming, gathering and managing household biomass fuels and water, and care-giving responsibilities) and their disproportionate lack of economic power, which is likely to have more serious ramifications in disaster situations. Wachholz (2007) also describes how the predicted rise in climate-related natural disasters is likely to be correlated with increases in violence against women within the regions that have experienced the extreme weather events (e.g., hurricanes, floods, and droughts) (see also Chap. 9).

Despite this diversity, there has been little *cultural criminological* attention to climate change. Building on broader investigations of cultural criminology–green criminology linkages (see, e.g., Brisman 2009, 2010a, b; Brisman and South 2010), as well as White's (2008, pp. 41–45) overview of "media reporting on the environment," this introductory and provisional chapter will begin to address the gap in the green criminological literature on climate change in two ways. First, it will present an argument for why it is important to consider media representations of environmental harm, in general, and climate change, in particular. Second, it will reflect on the meaning and significance of a specific example of media representations of climate change—the way in which the media has discussed (and has subsequently helped to perpetuate) climate change contrarianism.

I use the term "climate change contrarian" and "climate change contrarianism" rather than "climate change skeptic" and "climate change skepticism" because, as a number of scholars have pointed out, skepticism is a part of the scientific process.

Antilla (2005, p. 339n.5) acknowledges the difference between scientists who are skeptics "because the scientific process demands continuing questioning" and climate change skeptics who "have taken a highly visible public role in criticizing the scientific consensus on … climate change through publications and statements addressed more to the media and the public than to the scientific community," and then, having made this distinction, employs the term "climate skeptic" throughout her article. McCright and Dunlap (2003, p. 354n.8) state that "all scientists are (or at least should be) reflexive and skeptical." They explain that they use the term "climate change skeptics" instead of "skeptic scientists" in order "to convey the fact that what distinguishes … contrarian scientists from the vast majority in the scientific community is their strong and vocal dissent from the growing consensus regarding the reality of anthropogenic climate change" (McCright and Dunlap 2003, p. 354n.8). Anderegg (2010, p. 30), like McCright and Dunlap (2003), asserts that "[a]ll good scientists are skeptics, so the label climate change skeptic is not appropriate." But he also rejects such terms as "climate change contrarian" and "climate change denier," preferring instead to refer to such scientists as those who are "unconvinced by the evidence" (UE)—in contrast to those who are "convinced by the evidence" (CE). While I accept Anderegg's argument [as well as that of McCright and Dunlap (2003)] that "all good scientists are skeptics" and thus that the terms "climate change skeptic" and "climate change skepticism" may be problematic, I am unwilling to adopt his UE and CE terms because doing so would fail to acknowledge that many scientists who present themselves as "unconvinced by the evidence" receive funding from organizations and corporations that pay them to be *contrarians*—a point I discuss below. Thus, while *skepticism* can be both a healthy part of the scientific process and an excuse to present political or value-laden perspectives (that are masked behind a scientific façade), *contrarianism* suggests an ideological, rather than scientific, impetus for disagreement. And it is this element or characteristic that I wish to highlight.

The chapter begins with a brief introduction to cultural criminology and its concern with the mass media's representation of crime and criminal justice, as well as public conflicts over crime and crime control (see, e.g., Ferrell 1995, 1999, 2003). Taking inspiration from the cultural criminological position that "attention must be paid to the media and political dynamics underlying not only constructed crime but 'unconstructed' crime as well" [see Ferrell 2003, p. 74, citing Jenkins (1999); see also Jenkins (2009)], this chapter then argues that because green criminologists already pay attention to nonstatutorily proscribed environmental "harms" and "crimes" (see, e.g., Brisman 2008; Ruggiero and South 2010; White 2007, 2008), they may wish to analyze and assess the media and political dynamics surrounding the presentation of various environmental phenomena. Following Brisman and South's (2010) call for greater consideration of the "mediated construction of 'environment' and 'environmental harm'"—how concepts, terms, and real-world "environments," "landscapes," 'places," "spaces," and "wildernesses" are depicted in mainstream and popular media, as well as the ways in which environmental harms and disasters are reported, perceived, re-reported, re-perceived—and leaning on White's (2008, pp. 32–53) general discussion of the "social construction of environmental problems," this chapter next considers climate change contrarianism and contemplates what this might mean for collective and individual responses to this type of environmental harm.

Cultural Criminology and the Media

Cultural criminologists devote attention to popular culture constructions—especially mass media constructions—of crime and crime control, and they investigate how—and the ways in which—the dynamics of media and popular culture, the lives and activities of criminals and their subcultures, and the operations of social control and criminal justice converge in everyday life (see Ferrell 1999, pp. 395–396, Ferrell 2003, p. 71, Ferrell and Sanders 1995, p. 301, Ferrell and Websdale 1999, p. 3, Greer 2009, p. 177, 197; see also Hayward and Young 2004, p. 259, Hayward and Morrison 2009, p. 97, 98). Cultural criminologists consider "media images" to be "constructions, rather than reflections, of reality" (Ferrell and Sanders 1995, p. 308). As Ferrell (1995, p. 27) explains, "[t]o understand the reality of crime and criminalization … cultural criminology must account not only for the dynamics of criminal subcultures, but for the dynamics of the mass media as well. Today, mediated images of crime and criminal violence wash over us in wave after wave, and in so doing help shape public perceptions and policies in regard to crime." For cultural criminologists, the goal is not simply to expose the interconnections between the criminal justice system and the mass media (i.e., the ways in which the mass media relies on criminal justice sources for information about crime and images thereof) or to document the role of the media in shaping public perceptions of crime (see Ferrell 2003, p. 71, 73; Ferrell et al. 2008, p. 51, 123; see also Greer 2009, p. 177, 197). These are important to the cultural criminological endeavor, but cultural criminologists probe deeper and carry their inquiry further by investigating and dissecting the "*microcircuits of knowledge* regarding crime, deviance, and the societal reaction to these phenomena"—microcircuits that "often feed each other or blend into one another" (Websdale and Ferrell 1999, p. 349). According to Ferrell (1995, p. 34):

> criminal events, identities, and styles take life within a media-saturated environment, and thus exist from the start as moments in a mediated spiral of presentation and representation. Criminal events and public perceptions of criminality are reported on by the media less than they are constructed within the media; their existence is inevitably confirmed more by ratings points than by rates of crime. Criminal subcultures in turn reinvent mediated images as situated styles, but are at the same time themselves reinvented time and again as they are displayed within the daily swarm of mediated presentations. In every case, as cultural criminologists, we study not only images, but images of images, an infinite hall of mediated mirrors.

Cultural criminologists hold fast to the belief that "the reality of crime resides not in a particular action or event but in the flux of reactions and interactions through which the meaning of the action or event is negotiated and contested"

(Ferrell 2006, p. 248). As such, they attempt to understand not just the meaning of rule-breaking and transgression for participants in a subculture, but the media dynamics and other symbolic dimensions of crime and crime control, the politics of crime and crime control, and "the cultural channels through which these politics are played out" (Ferrell and Sanders 1995, p. 301; see also Ferrell et al. 2008, p. 85).

Cultural criminologists claim that mediated representations of crime affect individual and collective behavior, and thus have endeavored to understand "the emotions engendered by mediated images and collective representations of crime" (Ferrell et al. 2008, p. 71) and the ways in which these mediated characterizations of crime and crime control "shape attitudes and policy; define the effects of crime and criminal justice; generate fear, avoid and pleasure, and alter the lives of those involved" (Ferrell 2004, p. 4). One reason for cultural criminological attention to images of crime and crime control is that when subcultural activity—and subcultures, more generally—are criminalized, "they are primarily criminalized through the mass media, through their presentation and re-presentation as criminal in the realm of sound bites, shock images, news conferences, and newspaper headlines" (Ferrell 1999, p. 405). But cultural criminological attention to images of crime and crime control is also driven by a desire to know what mass media is *not* reporting on—what is *not* being depicted, why such stories are *not* being told, and what are the consequences of these decisions (see generally Brisman 2010/2011).

Green Criminology and the Media

Without making reference to the cultural criminological endeavor, a handful of scholars have begun to explore the environmental crimes, harms, and risks that the mass media is *not* reporting on—what the mass media is *not* depicting, why such stories are *not* being told, and what the consequences of these decisions are or might be. For example, Fitzgerald and Baralt (2010, p. 346) have observed that, "the media do not report on some risks. Whether or not a risk will be reported depends upon contextual issues, such as whether there are questions of blame, whether a visual image illustrating the impact of the risk is available, whether a large number of people are exposed to the risk and whether there is conflict among experts regarding the risk .... The attention paid to environmental risks varies depending upon how a given risk resonates with the larger culture [i.e., whether an environmental issue can be framed in a manner consistent with cultural givens]. ... [T]he media underestimates long-term risks from environmental toxins and pollutants, while more dramatic risks, such as accidents or natural disasters, are overestimated." (citing McCarthy et al. 2008; see also Tákacs-Sánta 2007, p. 31). Similarly, Simon (2000, p. 637) claims, "[o]ne reason why corporate crimes of all types flourish within certain industries is that the mainstream press underreports both individual incidents of crime and the seriousness of such violations. ... The lack of stories is also accompanied by vigorous public relations efforts that function to cover up environmental harms" (citations omitted). The position of Fitzgerald and Baralt, as well as that of Simon, is supported by Safina (2011), who documents how BP, in the aftermath of the Deepwater Horizon oil spill in the Gulf of Mexico in April 2010, posted private guards who closed off public beaches to keep journalists and witnesses away—leading Safina to observe that "[t]he flow BP is getting good at stopping is the flow of news" (Safina 2011, p. 127). In writing about a different type of environmental harm—climate change—Hulme (2009) makes an argument similar to that of Simon (as well as one that resonates with Fitzgerald and Baralt's comments about media coverage of risk and Safina's findings and BP). Hulme (2009, pp. 219, 227–228) asserts that "the media do not operate as a neutral conveyor of scientific knowledge to a passive audience. They actively and continuously engage in framing, filtering and interpreting messages about climate change using affective and emotive language and imagery. ... By definition, frames select and emphasize certain facets of an issue and must therefore de-select and de-emphasize others. Communicating climate change can therefore never be merely 'raising awareness' or simply presenting 'the facts.'"

Despite the fact that the mainstream press underreports environmental risks, individual incidents of environmental crime and harm, and the seriousness of such violations, green criminologists have largely neglected the study of the mediated representation or construction of "environmental crime" and "environmental harm," and the ways in which public understanding is shaped by these media representations or constructions. Despite the fact that many environmental battles are "won" or "lost" in the press (compare Castells 2004, pp. 186–187, Doyle 2007, Smith 2005, p. 1473, and Turner 1993 and with Boekhout van Solinge 2010, p. 271) and that according to some, direct confrontation and the media are "essential qualities" of today's environmentalism—especially today's radical environmentalism (Scarce 1990, p. 48; see also Castells 2004, pp. 186–187; McKay 1998, p. 9, 11)—green criminologists have actually devoted relatively little attention to the media and political dynamics surrounding the presentation of various environmental phenomena.

There are a few exceptions worth noting. Lynch et al. (1989) have examined media constructions of the Bhopal disaster, and Lynch et al. (2000) have studied media portrayals of chemical crimes in Tampa, Florida. Yates et al. (2001) investigate the moral panic associated with a series of assaults on horses in England from the early 1990s onward, and explore "how the nature of the relationships between humans and animals other than humans ... is revealed through authoritative utterances [about offenders and victims] by the media"; they suggest that "understanding how and when victimhood is ascribed to animals helps uncover the invisible assaults that are routinely inflicted on animals and against whose perpetrators the categories of criminalization are almost never applied" (Yates et al. 2001, p. 143, 159). More recently, Jancell (2007) has contemplated media representations of federal petroleum refining violations, and Fitzgerald and Baralt (2010) has explored the case of mercury-contaminated fish as depicted in *The New York Times* and the *Globe and Mail* from 2003 through 2008.

These examples notwithstanding, there has been little concerted effort among green criminologists to examine the dynamics surrounding the mediated representation or construction of "environmental crime," "environmental harm," and "environmental risk," and the ways in which public understanding is shaped by these media representations or constructions. This is unfortunate because just as "the mass media provide the preponderance of the public's information on crime and crime control" (Ferrell 2009, p. 169), the mass media now provide the preponderance of the public's information on "environment" and "nature" (see, e.g., Antilla 2005, p. 339; Boykoff and Boykoff 2004, p. 125; Carvalho 2007, p. 223; Dietz et al. 2007, p. 190, 209; Hansen 1991; Nelkin 1995; Whitmarsh 2009, p. 405; cf. Tákacs-Sánta 2007, p. 29, 30)—as well as the preponderance of the public's information on environmental harm (see, e.g., Schoenfeld et al. 1979; Spector and Kitsuse 1977), such as the problem of mercury-contaminated fish (see Fitzgerald and Baralt 2010, pp. 342–344), global climate change (see, e.g., Antilla 2005, p. 339, 350; Wilson 1995, 2000a, b), and the Deepwater Horizon oil spill. In addition, just as the media (newspapers, television, radio, and the Internet) tell a story of *crime* that we store away in our consciousness to be used when we make decisions in our everyday lives, they also tell a story about nature, environment, and environmental harm that we store away in our consciousness to be used when we make decisions in our everyday lives.

Green criminologists should scrutinize the information provided by the media—especially, the ways in which the media depict (or do not depict) the environmental risks, the causes of environmental harm, and the responsibility for mitigating environmental harm (e.g., governmental regulation of industry, industry itself, individuals). Green criminologists might also study the "stories" told about environment, nature, and environmental harm, and the linkages between these "stories" and decision-making and behavior—for, as Fitzgerald and Baralt (2010, p. 345) note, government discourses often have a strong effect on the media, and the media, in turn, "can have an agenda-setting effect and influence which environmental issues the public will be concerned about" (see also Carvalho 2005, 2007; Hansen 1991). Cultural criminology provides a model, but whereas cultural criminologists frequently examine various media of mass communication in order to understand public support for particular crime and crime control policies, or public concerns over everyday criminality, green criminologists may wish to examine various media of mass communication in order to understand (the reasons behind) *the absence* of public support for particular environmental protection strategies or *lack* of concern over macro- and microlevel environmental catastrophes.

It is this absence of support for national and international climate change legislation to which I turn next. Although my focus is primarily on the absence of public and political support for international and national climate change legislation in the United States, I draw on media from around the world in making my arguments, and my hypotheses regarding public and political apathy and inertia may be relevant in other countries. In fact, scholars such as Jacques et al. (2008, pp. 360–361; see also McCright and Dunlap 2003, p. 370) suggest that conservative think tanks in the United States have helped diffuse "environmental skepticism" internationally.

Thus, an examination of *the absence* of public support in the United States for particular environmental protection strategies or *lack* of concern in the United States over macro- and microlevel environmental catastrophes—as well as the role of corporate–political powers and the media in contributing to such absence of public support and lack of concern—may well shed light on similar phenomena in other countries.

Climate Change Contrarianism and the Media

In February 2007, William K. Stevens, the lead reporter for *The New York Times* on climate change until his retirement in 2000, wrote: "one day, if mainstream scientists were right about what was going on with the earth's climate, it would become so obvious that human activity was responsible for a continuing rise in average global temperature that no other explanation would be plausible. That day may have arrived." According to Stevens (2007, p. D3), "the steadily strengthening body of evidence about the human connection with global warming is at least approaching the higher standard [beyond a reasonable doubt—the standard required in most criminal cases—rather than preponderance of the evidence—the standard for most civil cases] and may have already satisfied it." Stevens (2007, p. D3) reasoned that "a demonstrably heightened awareness and concern among Americans about global warming ... had been energized largely by dramatic reports on the melting Arctic and by fear—generated by the spectacular horror of Hurricane Katrina—that a warmer ocean is making hurricanes more intense."

While glaciers and polar icecaps have continued to recede, belief that climate change is "real" and confidence in climatic science has surprisingly decreased as well (see, e.g., Broder 2010, Editorial 2010b, d; Friedman 2010b; Kaufman 2010a, b). In 2010, the annual Gallup Social Series Environmental poll revealed that the American public had become less worried about the threat of climate change, less convinced that its effects were already happening, and more likely to believe that scientists themselves were uncertain about its occurrence—with 48% of Americans' surveyed expressing the belief that the seriousness of climate change is generally exaggerated, up from 41% in 2009 and 31% in 1997 (Newport 2010). Although citizens of other developed countries tend to express greater belief than Americans that humans are causing climate change (see, e.g., Anderegg 2010, p. 27; Dietz et al. 2007, p. 186), growing doubt, disbelief, and denial are not just endemic to the United States. Angus Reid polls conducted in December 2009 found declining support for climate change as a human-induced phenomenon in Britain, Canada, and the United States [although the drop-offs from July and November 2009 to December 2009 might be attributed, in part, to the failure of the United Nations Climate Change Conference in Copenhagen in December 2009 to generate a legal binding treaty to reduce CO_2 emissions (see Corcoran 2010)]. In February 2010, the BBC found that only 26% of Britons believed that "climate change is happening and is now established as largely manmade"—down from 41% in November 2009; a poll conducted for the German magazine, *Der Spiegel*, found that 42% of Germans feared global warming—down twenty percent from 4 years earlier (Rosenthal 2010a).

To some extent, there may always be some doubt and disbelief about climate change. As Stevens (2007, p. D3) himself noted, "[m]any gaps remain in knowledge about the climate system. Scientists do make mistakes, and in any case science continually evolves and changes." Some will hold fast to the belief "climatologists do not know enough about long-range patterns to draw definitive links between global warming and weather extremes" (Gillis 2010c) or that global warming can be explained by "natural earth cycles" (Kaufman 2010b). But, as the sociologist William Foote Whyte once remarked, "humans would take very few actions if they only responded to scientifically tested propositions" (1993, p. 370). Or, as Gleick et al. (2010, p. 689) explain, "[t]here is always some uncertainty associated with scientific conclusions; science never absolutely proves anything. When someone says that society should wait until scientists are absolutely certain before taking any action, it is the same as saying society should never take action." Furthermore, as Wachholz (2007, p. 178) explicates, "[c]limate change has typically been seen as a technical problem that will be solved through technical solutions. It must also be understood as a *social process* ...." (emphasis added). For some, disbelief may stem from religious conviction that humans could not possibly alter a world created by God or that God created the Earth for humans to (ab)use (see, e.g., Kaufman 2010a, b; Broder 2010; see generally Moser and Dilling 2004, pp. 35–36). For others, uncertainty may result from the lack of a readily identifiable villain (Hulme 2009, p. xxi), difficulty "seeing" the effects or impacts of climate change, difficulty of connecting products people purchase to climate change (Marx 2008; see also Lowe et al. 2006, p. 437), and misunderstanding of expert knowledge—or, at least, incongruities in the ways in which we trust expert opinion and manage risk (Anderegg 2010; Dietz et al. 2007, pp. 189–190, 208).

Though there appears to be widespread agreement among scientists that climate change is occurring and that human activities are probably driving it (see, e.g., Anderegg 2010; Begley 2007; Boykoff and Boykoff 2004, p. 125, 129; Carvalho 2007, p. 223, 232; Gleick et al. 2010; Kakutani 2010; Lynch et al. 2010, p. 229, 234; McCright and Dunlap 2003, pp. 348–350, 366; Ockwell et al. 2009, p. 305; Whitmarsh 2009, p. 401; cf. Tierney 2009), many Americans, including many American politicians and decision-makers, are increasingly viewing climate change as a "left-wing plot" (Editorial 2010e)—part of the "one-world socialist agenda" (http://www.climategate.com/about) or a "conspiracy to impose world government and a sweeping redistribution of wealth" (Broder 2010). Just as Republican Senator James Inhofe of Oklahoma proclaimed on the Senate floor that "[g]lobal warming is the greatest hoax ever perpetrated on the American people" (cited in Antilla 2005, p. 338, 347), many Americans believe that climate change is "a cynical hoax perpetrated by climate scientists ... greedy for grants" (Kaufman 2010b; see also Anderegg 2010, p. 23; Douthat 2010, p. A23; Editorial 2010b, c, d; Friedman 2010c, p. WK8; Warner 2011, p. 11). A number of factors have contributed to this changing tide—to the increasing belief that the seriousness of climate change is generally exaggerated and to the diminishing conviction of and need to respond to anthropocentric influence on the global climate system. I consider these in turn.

The first factor might be considered the "residual effects" of the Bush Administration's 8-year-long process of systematically ignoring and denying scientific evidence of climate change, while simultaneously promoting national and international environmental policies beneficial to private economic and industry interests—many of whom had backed Bush's election campaign (see, e.g., Doyle 2007, p. 143). As exhaustively demonstrated by Lynch et al. (2010, pp. 222–226), the Bush Administration hid from public view scientific studies and documents with which it disagreed (including censuring US Environmental Protection Agency climate change documents); required all federal agencies intending to issue press releases or reports on climate change to submit such items to the White House for prior review; attempted to limit media access to federal climate scientists; manipulated government reports on climate change; and pressured government climate scientists to delete references to "global warming" or "climate change" from government-sponsored papers and reports. In addition to these repeated and pervasive attempts by the administration of George W. Bush to suppress climate change science (see, e.g., Easterbrook 2011), the Bush Administration refused to sign the Kyoto Protocol (or negotiate any mandatory global warming gas-reduction agreements, for that matter); failed to meet its obligations under the Global Change Research Act (which requires periodic reports on the effects of and projections for global climate change); and relaxed key components of the Clean Air Act's New Source Review rules, thereby allowing owners of the country's oldest and dirtiest power plants, as well as oil refineries, chemical plants, incinerators, iron and steel foundries, pulp and paper mills, cement plants and other factories and industrial facilities to make upgrades without installing pollution controls (see Lynch et al. 2010, pp. 221–222, 225–226; see also Brisman 2005, pp. 63–66; Doyle 2007, p. 143). Taken as a whole, we can point to the Bush Administration's refusal to accept, much less develop and act upon climate change science, as a reason for low public and political concern for climate change; current economic conditions, ongoing wars in Afghanistan and Iraq, and President Obama's privileging of health care legislation over climate change policy (see Lizza 2010, p. 75; Kerr 2009, p. 928; see generally Jacques et al. 2008, p. 351, 362) have not helped to reverse course.

The second factor might be attributed to admitted errors by the United Nations Intergovernmental Panel on Climate Change (IPCC)—the intergovernmental body charged with reviewing and assessing international scientific, technical, and socio-economic information on climate change (including current and potential environmental and socioeconomic impacts of human-induced climate change, and options for adaptation and mitigation). The Fourth Assessment Report (AR4), which was published in 2007—the same year that it shared the Nobel Peace Prize with former Vice President Al Gore—included some of the following key conclusions:

- Warming of the climate system is unequivocal, as is now evident from observations of increases in global average air and ocean temperatures, widespread melting of snow and ice and rising global average sea level.
- Observational evidence from all continents and most oceans shows that many natural systems are being affected by regional climate changes, particularly temperature increases.

- Other effects of regional climate changes on natural and human environments are emerging, although many are difficult to discern due to adaptation and non-climactic drivers.
- Global GHG emissions due to human activities have grown since pre-industrial times, with an increase of 70% between 1970 and 2004.
- There is very high confidence that the global average net effect of human activities since 1750 has been one of warming.
- Most of the observed increase in global average temperatures since the mid-twentieth century is very likely due to the observed increase in anthropogenic GHG concentrations.
- It is likely that there has been significant anthropogenic warming over the past 50 years averaged over each continent.
- Anthropogenic warming over the last three decades has likely had a discernible influence at the global scale on observed changes in many physical and biological systems.
- There is high agreement and much evidence that with current climate change mitigation policies and related sustainable development practices, global GHG emissions will continue to grow over the next few decades.
- Continued GHG emissions at or above current rates would cause further warming and induce many changes in the global climate system during the twenty-first century that would very likely be larger than those observed during the twentieth century.
- Anthropogenic warming and sea level rise would continue for centuries due to the time scales associated with climate processes and feedbacks, even if GHG concentrations were to be stabilized.
- Anthropogenic warming could lead to some impacts that are abrupt or irreversible, depending upon the rate and magnitude of the climate change.
- Adaptation can reduce vulnerability, both in the short and the long term.
- [T]here is high agreement and much evidence of substantial economic potential for the mitigation of global GHG emissions over the coming decades that could offset the projected growth of global emissions or reduce emissions below current levels.
- There is also high agreement and medium evidence that changes in lifestyle and behavior patterns can contribute to climate change mitigation across all sectors.

Despite these conclusions, public attention surrounded small errors in the report. For example, in the second paragraph of section 10.6.2 of page 493 of the 900-plus-page report of the Contribution of Working Group II to the Fourth Assessment Report of the Intergovernmental Panel on Climate Change states:

> Glaciers in the Himalaya are receding faster than in any other part of the world … and, if the present rate continues, the likelihood of them disappearing by the year 2035 and perhaps sooner is very high if the Earth keeps warming at the current rate. Its total area will likely shrink from the present 500,000 to 100,000 km^2 by the year 2035.

The better estimate is that Himalayan glaciers will disappear by *2350*, not *2035*, but the mistake was reported as "wildly inaccurate," as an "egregious error," and as an instance of "apparent bias" (Bagla 2009; Webster 2010). A *BBC News* headline read "Himalayan glaciers melting deadline 'a mistake'" (Bagla 2009), while the conservative British paper, *The Times*, ran a report, "UN must investigate warming 'bias', says former climate chief" (Webster 2010).

In the United States, *The New York Post*, in an article titled "The EPA's climate con" (22 February 2010), reported that a "key finding" of AR4 was "utterly bogus," and asserted that the IPCC had "*no* evidence" for its claims regarding the Himalayan glaciers (emphasis in the original). This article began with the line "[l]ooks like the great climate-change unraveling came none too soon," and reported that "climate change [a]larmists have been reduced to arguing that drastic steps should be taken to curtail carbon emissions *just in case* climate change is as bad as they say ... an argument [that] completely ignores the costs of such a move" (emphasis in the original).

A subsequent article by *The New York Post*, bearing the headline "Meltdown of the climate 'consensus'" (Patterson 2010), began: "If this keeps up, no one's going to trust *any* scientists" (emphasis in the original). The same article, which was written by an editor of a publication produced by the Capital Researcher Center—a conservative, nonprofit organization that promotes "private alternatives to government regulatory and entitlement programs"(http://www.capitalsearch.org/about/) and that "discourage corporations from giving charitable donations to nonprofits that support liberal or anti-business policies" (Barr 1997)—ended by stating that the claims of "warming alarmists" are "now in shreds" and that "the best evidence" tells us "[t]hat man-made global warming is a mere hypothesis that has been inflated by both exaggeration and downright malfeasance" (Patterson 2010).

The New York Times reported that the 2035 date "ha[d] been much quoted and a cause for enormous consternation" (Rosenthal 2010b, p. A8). But the article, which contained the less inflammatory title, "U.N. Panel's Glacier Warning Is Criticized as Exaggerated," made clear that "the potentially erroneous figure in question had appeared only in the panel's full report of more than 1,000 pages and had been omitted in later summary documents that the panel produced to guide policy," and observed that error was "the latest in a string of events that climate change skeptics have seized on to support their contention that fears about warming are unfounded, or at least overblown" (Rosenthal 2010b, p. A8).

Dr. Martin Parry, the cochair of the IPCC's working group on impacts (Working Group II: Impacts, Adaptation and Vulnerability), which produced the report that included the incorrect claim about the melting of the Himalayan glaciers, expressed surprise and puzzlement at the way in which the media had focused on minor points: "What began with a single unfortunate error over Himalayan glaciers has become a clamor without substance" (quoted in Adam 2010). But Dr. Rajendra Kumar Pachauri, chair of the IPCC, confronted critics of the IPCC and AR4, and expressed regret for the Himalayan glacier error (see Bagla 2010, pp. 510–511). The IPCC also issued a statement "regret[ting] the poor application of well-established IPCC procedures," while reasserting its belief that its overall conclusions about global and regional climate change were "robust, appropriate, and entirely consistent with the underlying science" (IPCC 2010). Nevertheless, and as alluded to above, various corporate–political interests jumped upon the small errors in the 2007 report from the IPCC to suggest that the whole climate change phenomena was a "hoax" (Editorial 2010b; see also Anderegg 2010, p. 23; Douthat 2010, p. A23; Editorial 2010d; Friedman 2010c, p. WK8; Warner 2011, p. 11). As Carvalho (2007, p. 237) explains, "[w]hen knowledge claims appear[] to constitute a threat to ideological principles and arrangements in the political, social and economic realms, [corporate–political entities] do not hesitate to harm the reputation of an institution like the IPCC"—a reputation that has yet to recover.

The third factor that might be contributing to decreasing perceptions of the seriousness of global climate change and the need for individual and collective action is the Climate Research Unit (CRU) email controversy—which has been dubbed "Climategate" (see, e.g., Revkin 2009b; Sussman 2010). In November 2009, thousands of emails and other documents from the University of East Anglia's (UEA) Climate Research Unit (CRU) were made public, after being illegally obtained through hacking of the UEA's computers—documents that climate change doubters, deniers, and disbelievers claimed proved that mainstream climate scientists were overstating the case for human influence on climate change (see, e.g., Editorial 2009; Kintisch 2009; Revkin 2009a). Subsequent investigations, however, concluded that UEA and CRU-affiliated climate science researchers had not colluded to withhold scientific information, interfered with the peer-review process to prevent dissenting scientific papers from being published, deleted raw data, or manipulated data to make the case for climate change more compelling (see Associated Press 2010; Editorial 2010a; Gillis 2010a, b; Rosenthal 2010a; see also Krugman 2010b). Dr. Phil Jones, Director of the CRU, and Dr. Michael E. Mann—a climatologist at Pennsylvania State University, who has been singled out over the years by climate change deniers for his research that demonstrates that the Earth's temperatures have risen steadily since the early 1990s—were among those cleared of allegations of wrongdoing (Begley 2010). Dr. Mann, who worked at the University of Virginia from 1999 to 2005, has also withstood legal challenges brought by Ken T. Cuccinelli, the Attorney General of Virginia and a conservative Republican who is doubtful about climate change (Associated Press 2010; Helderman 2010). (Cuccinelli has accused Mann of violating the state's Fraud Against Taxpayers Act by using manipulated data to obtain grants, but in late-August 2010, the judge presiding over the case ruled that Cuccinelli had failed to state an objective reason to believe that Mann had committed fraud when he sought public grants to fund his research).

In the aftermath, Begley (2010) described "Climategate" as "that *highly orchestrated, manufactured scandal*, [where] e-mails hacked from computers at the University of East Anglia's climate-research group were spread around the Web by activists who deny that human activity is altering the world's climate in a dangerous way, and spun so as to suggest that the scientists had been lying, cheating, and generally cooking the books" (emphasis added). Similarly, Krugman (2010b) stated: "You've probably heard about the accusations leveled against climate researchers—allegations of fabricated data, the supposedly damning e-mail messages of 'Climategate,' and so on. What you may not have heard, because it has received much less publicity, is that everyone of these supposed scandals was eventually unmasked as a fraud concocted by opponents of climate action, then bought into by many in the media." And an Editorial in *The New York Times* expressed hope that "[p]erhaps now we can put the manufactured controversy known as Climategate behind us and turn to the task of actually doing something about global warming" (Krugman 2010b).

Despite the pronouncement that "[a]nother (no less overblown) climate change controversy may also be receding from view" (Editorial 2010c)—the other controversy being the one surrounding the 2035 date in the IPCC's AR4—some claim that Climategate has damaged the public credibility of climate research, while emboldening doubters and deniers of climate change (see generally Kintisch). The Climategate ruckus has also been identified as being responsible for some polls finding decreased support for the proposition that climate change is underway (see, e.g., Barringer 2011; Corcoran 2010).

While one could point to the residual effects of the cooperative corporate-state anti-climate change strategy under the George W. Bush Administration as reasons for diminished belief in the reality of climate change, decreased concern for climate change, and reduced support for action to curb or mitigate climate change, George W. Bush has been out of office since January 2009. The IPCC has moved past the ballyhoo surrounding errors in its AR4 and has been working on the Fifth Assessment Report (AR5), which will be completed in 2014. And, as noted above, the UEA- and CRU-affiliated climate science researchers were cleared of wrongdoing—there was no "conspiracy to manipulate research to support predetermined ideas about global warming" (Editorial 2010c). In addition, according to annual surveys of California's environmental attitudes by the Public Policy Institute of California, while the Climategate fiasco seemed to be responsible for knocking 7 percentage points off the total share of Californians who in 2010 thought climate change was underway, the seven points were regained in 2011 (see Barringer 2011). Thus, whatever negative impact the AR4 scandal and Climategate hubbub might have had should now have dissipated. In fact, two other major scientific reports have produced disquieting findings and dire warnings about global climate change, which should have restored belief in the reality of climate change, renewed concern for climate change, and reinvigorated support for action to curb or mitigate climate change.

In September 2009, the United Nations Environment Programme (UNEP) issued a report, entitled "Climate Change Science Compendium 2009," reviewing "some 400 major scientific contributions to our understanding of Earth Systems and climate that have been released through peer-reviewed literature or from research institutions over the last 3 years, since the close of research for consideration by the IPCC Fourth Assessment Report" (http://www.unep.org/compendium2009/). Although the UNEP report was "not the exhaustively peer-reviewed consensus assessment of peer-reviewed literature that the IPCC produces every 5 or 6 years" (Kerr 2009, p. 927), the "Climate Change Science Compendium 2009" contained the latest peer-reviewed scientific findings, interpretations, ideas, and conclusions. These findings were, in many ways, more sobering than those in IPCC's AR4, leaving Ban Ki-moon, Secretary-General of the United Nations, to state in the Foreword to the report: "The science has become more irrevocable than ever: Climate change is happening. The evidence is all around us. And unless we act, we will see catastrophic consequences including rising sea levels, droughts and famine, and the loss of up to a third of the world's plant and animal species. ... [C]limate change is accelerating at a much faster pace than was previously thought by scientists. ... The time for hesitation is over" (United Nations Environment Programme 2009).

More recently, in May 2011, the National Research Council, an arm of the National Research Council, issued a report, "America's Climate Choices," stating that "the environmental, economic, and humanitarian risks of climate change indicate a pressing need for substantial actions to limit the magnitude of climate change and to prepare for adapting to its impacts," judging that "the risks associated with doing business as usual are a much greater concern than the risks associated with engaging in ambitious but measured response efforts," and urging the federal government to "immediately undertake the development of a national adaptation strategy and build durable institutions to implement that strategy and improve it over time" (National Research Council et al. 2011, p. 27, 36, 66). The National Research Council report also recommended that "[t]he United States should actively engage in international level climate change response efforts: to reduce greenhouse gas emissions through cooperative technology development and sharing of expertise, to enhance adaptive capabilities (particularly among developing nations that lack the needed resources), and to advance the research and observations necessary to better understand the causes and effects of climate change" (National Research Council et al. 2011, p. 74).

Given such increasingly clear and authoritative assessments of climate change, something else must be at play. Something other than—or more than—the lingering effects of the George W. Bush's administration's misdeeds and overblown Himalayan glacier and Climategate controversies must be contributing to the ongoing American refusal to accept and act upon the overwhelming scientific evidence that climate change is occurring and that human activities are probably causing it—a far cry from the sentiments Stevens conveyed in early 2007 about the "demonstrably heightened awareness and concern among Americans about global warming."

U.S. quiescence is not just the result of the enduring impact of the 8-year collusion between the Bush Administration and the corporate sector. Nor can we treat the inflated IPCC and Climategate controversies as discrete events. Rather, US inaction on climate change has stemmed, in large part, from various corporate–political interests who, wishing to downplay the extent or existence of climate change, have conducted a concerted campaign to try to call the science behind the phenomena into question. The above examples are evidence of that resolute battle—of fossil fuel industry-funded climate change naysayers' and contrarians' relentless commitment to raising doubts about the science of climate change and to undermining policies devised to address it (see, e.g., Beder 1999; Begley 2007; Boykoff and Boykoff 2004; Carter 2007; Doyle 2007; Gelbspan 1998, 2004, 2005; Jacques et al. 2008; Leggett 2001; McCright and Dunlap 2000, 2003; Weart 2003; see generally Tákacs-Sánta 2007, p. 31). Because the restraints on production and consumption of greenhouse gases would cut against the interests of key business sectors (see, e.g., Jacques et al. 2008, p. 354; McCright and Dunlap 2000, p. 504, 505, McCright and Dunlap 2003, p. 353; Tákacs-Sánta 2007, p. 31), such "economic elites," to use Benton's (2007, p. 8) term, have "created and lavishly financed institutes to produce anti-global-warming studies, paid for rallies and Web sites to question the science, and generated scores of economic analyses that purport to show that policies to reduce emission of climate-altering gases will have a devastating effect on jobs and the overall economy" (Broder 2010, p. A4)—a perspective shared by Friedman (2010c, p. WK8), who contends that "the fossil energy companies, driven by the need to protect hundreds of billions of dollars of profits, encourage obfuscation of the inconvenient scientific results." Similarly, Krugman (2010a, p. A23) has opined: "If you want to understand opposition to climate action, follow the money. ... Look at the scientists who question the consensus on climate change; look at the organizations pushing fake scandals; look at the think tanks claiming that any effort to limit emissions would cripple the economy. Again and again, you'll find that they're on the receiving end of a pipeline of funding that starts with big energy companies, like Exxon Mobil, which has spent tens of millions of dollars promoting climate-change denial, or Koch Industries, which has been sponsoring anti-environmental organizations for two decades."

Krugman's advice to "follow the money" is quite sound. As Antilla (2005, p. 350) explains, "[t]here have *always* been experts willing to back up a 'profitably mistaken viewpoint'; there have always been efforts 'to cover the issue in a thick fog of sophistry and uncertainty' and to 'unearth yet one more reason why the status quo is best for us'" (quoting Nissani 1999, pp. 37–38 (emphasis added)). When ExxonMobil [which provided funding for Bush's election campaign (Doyle 2007, p. 143)] offers scientists five-figure rewards to write articles that undercut reports demonstrating the reality of anthropogenic climate change (see Begley 2007; Lynch et al. 2010, p. 228; Schulman et al. 2007; see also http://www.exxonsecrets.org/maps.php)—or when the company donates millions of dollars to conservative think tanks and organizations (such as the International Policy Network, the George C. Marshall Institute, the Competitive Enterprise Institute, and the Scientific Alliance) that dispute the impact of climate change and that emphasize the uncertainties in/of climate science (see, e.g., Adam 2006; Barnett and Townsend 2004; Boykoff and Boykoff 2004, p. 133; Lynch et al. 2010, p. 228)—this is no different from the tobacco industry's denials of the link between lung cancer and smoking (Begley 2007; Jacques et al. 2008, pp. 361–362; Lynch et al. 2010, p. 227) or "previous assaults on science, such as by the pesticide industry (DDT), coal-burning electric utilities (acid rain), and the chemical industry (effects of CFCs on stratospheric ozone)" (Antilla 2005, p. 350 citing Pollack 2003). This "phoney" controversy surrounding human-caused climate change "has been preceded by controversies on such issues as slavery, child labor, and civil rights" (Antilla 2005, p. 350, quoting Nissani 1999, p. 37).

As stated in the previous paragraph, the fossil fuel industry has funded both conservative think tanks who publicly challenge what they perceive as the false consensus of "mainstream" climate science and contrarian scientists willing to dissent from this growing consensus, as well as the political campaigns of many politicians sympathetic to (or who quickly become sympathetic to) fossil fuel-industry needs. To illustrate, when the National Research Council issued "America's Climate Choices," Representative Joe L. Barton (R-Tex.)—a leading opponent of regulating carbon emissions—dismissed the report: "I see nothing substantive in this report that adds to the knowledge base necessary to make an informed decision about what steps—if any—should be taken to address climate change" (quoted in Kaufman 2011, p. A16). Since 1997, Barton, who rejects the existence of climate change, has raked in millions of dollars in campaign contributions from oil, gas, and coal industries, as well as electric utilities (see, e.g., Crowley 2006; see also Calmes 2010b). According to the nonpartisan Center for Responsive Politics, the top two industries contributing to his campaign committees have been electric utilities and oil and gas industries (http://www.opensecrets.org/politicians/summary.php?cid=N00005656&cycle=2010). This may help to explain why, in 2005, Barton (as chairman of the House Energy and Commerce Committee) launched an investigation and demanded personal and private information from scientists who had coauthored a 1999 study demonstrating dramatic increases in climate change over the past millennium (Editorial 2005)—or why, in June 2010, he accused the White House of a "$20 billion shakedown" of BP after the company agreed to establish a liability fund to help those affected by the Deepwater Horizon oil spill in the Gulf of Mexico (see, e.g., Becker and Hulse 2010; Calmes 2010a). Barton also referred to President Obama's announcement about the liability fund as "a tragedy of the first degree" (quoted in Becker and Hulse 2010)—an especially perverse choice of words and a further indication of his intimate relationship with "Big Oil."

While special-interests' campaign contributions will often affect governmental policy in a system of representational democracy, corporate funding of scientists, organizations, and think tanks for the express purpose of undermining climate science and sowing dissent is far more insidious, invidious, and pernicious. These climate change contrarians operate outside the scientific community. Begley (2007) reports that the "denial machine" is able to reinforce the appearance of uncertainty and disagreement by churning out white papers and "studies" that are critiques of others' work, rather than empirical research. According to McCright (2007, p. 202), although "[s]ome of the contrarians publish in the peer-reviewed climate science literature, where they oversimply and even misinterpret existing research while selectively presenting data supporting their own counterclaims … most contrarians challenge climate change knowledge claims largely through activities outside of the scientific community" (internal citations omitted). McCright (2007, p. 202) continues:

> For presenting their most dubious assertions, they have chosen venues that are free from the constraints of traditional scientific standards. This withdrawal from the institutions and processes that define modern science provides the contrarians with great latitude in making their arguments. For instance, most contrarians present claims that consistently exceed the content of their peer-reviewed work in publications, public appearances, and websites supported by fossil fuels organizations and conservative think tanks. And the contrarians make their assertions to lay audiences who may not detect the technical flaws in their arguments. … Since their credentials inspire perceptions of expertise and trustworthiness

> among non-experts, the lack of accountability outside the scientific community makes the contrarians especially dangerous to scientific communication efforts. They can present assertions that do not withstand scientific peer review to an audience that often assumes, because of the contrarians' credentials, that those arguments are sound and constitute scientific evidence (internal citations omitted).

Operating unencumbered by peer-review, climate change contrarians are able not only to deny the existence of anthropocentric climate change or to belittle its effects, but have been able to step out of the realm of science and weigh in on issues well-outside their realm of purported expertise (see generally Jacques et al. 2008, p. 356). According to Moser and Dilling (2004, p. 38box), "[c]ommon contrarian tactics—tried and honed in previous anti-environmental and anti-consumerist safety campaigns—go through a well-known sequence: denying the problem, downplaying its severity, predicting economic ruin, and relying on human adaptive capacity and ingenuity. All along the way, the proponents of these views exploit scientific uncertainties, use selective decontextualized scientific findings, call on flawed pseudoscientific studies, and bank on the ignorance of the general public to support their views, while peppering their public statements with derogatory name-calling and portrayals of scientists and politicians." McCright and Dunlap (2000) have also found climate change contrarians to couple criticism of mainstream climate science with economic prognostications. They analyzed the claims made and framing techniques promoted by certain US conservative think tanks in their attempts to dismiss the reality of climate change, and, focusing on the years 1990–1997, identified three major counterclaims—(1) criticism of scientific evidence for global climate change—i.e., that the problematic condition does not exist; (2) arguments for substantial benefits of climate change; and (3) threats to the US economy and sovereignty if actions to reduce or mitigate climate change were undertaken.

While climate change contrarians benefit from the financial support of the fossil fuel industry and enjoy a certain level of political access (McCright 2007, p. 200, 202, 204; McCright and Dunlap 2003, p. 357), the question remains: how have climate change contrarians, who are operating outside the scientific community, garnered so much attention and succeed in influencing politicians and public opinion? In other words, while full financial coffers and ease of political access can keep many misguided ideas, viewpoints, and policy proposals afloat, no amount of money or political entrée could sustain, for example, claims that the Earth is flat. How then are climate change contrarians able to stay in the public eye and confuse both policy-makers and the general public about climate change—especially with growing international scientific consensus on anthropogenic climate change?

According to McCright (2007, p. 203), climate change contrarians exploit the media's balancing norm—the media's balanced presentation of "pro" and "skeptical" climate change scientists (see also McCright and Dunlap 2003, p. 366). Lowe et al. (2006, p. 436) state that the people in the United States "have been misled by newspaper reports that tend give equal weight to both sides of the climate change debate" and that "[t]he journalistic practice of balancing the scientific consensus with a comparatively small number of contrarians has acted to overstate the actual degree of disagreement." Likewise, Moser and Dilling (2004, p. 36) explain that "the media … tends to portray the climate change issue as one of large uncertainty, filled with competing claims and intends debate within the scientific community. The common practice of giving equal time to unequal sides is highly misleading. 'Balancing' the scientific consensus with the voices of a comparatively tiny number of contrarians overstates the actual degree of disagreement. This reinforces the public's perception of uncertainty and adds to confusion" (see also White, 2008).

Such perspectives on the dynamics of media representation of climate change are supported by the research of Boykoff and Boykoff (2004) and Antilla (2005). Boykoff and Boykoff analyzed a sample of 636 news articles in *The New York Times*, *The Washington Post*, *Los Angeles Times*, and *The Wall Street Journal* between 1988 and 2002 and found a significant difference between scientific views of climate change (although they used the search term, "global warming") and media presentations about climate change—specifically, a significant difference between the scientific community discourse and the US prestige-press discourse regarding the existence of anthropogenic contributions to climate change and decisions regarding action on climate change. They concluded:

> Focusing attention on "both" sides of the story regarding action due to global warming, the US prestige press in effect provided "balanced" coverage of a very unbalanced issue.
>
> …
>
> Through overwhelmingly "balanced coverage" of various decisions regarding action due to global warming, the prestige press thereby implied that the division between various calls for action was relatively even. In light of the general agreement in the international scientific community that mandatory and immediate action is needed to combat global warming, US prestige-press coverage has been seriously and systematically deficient. … By empirically unpacking the robust norm of balanced reporting, this research examines what may on the surface be an obvious journalistic tendency—the proclivity to tell "both sides of the story"—and excavates it to find that balanced reporting is actually problematic in practice when discussing the human contribution to global warming and resulting calls for action to combat it.
>
> …
>
> [T]he US prestige press—the New York Times, the *Washington Post*, the *Los Angeles Times*, and the *Wall Street Journal*—has contributed in significant ways to this failed discursive translation through the adherence to journalistic norms, and more specifically to the journalistic norm of balance. In the end, adherence to the norm of balanced reporting leads to informationally biased coverage of global warming. This bias, hidden behind the veil of journalistic balance, creates both discursive and real political space for the US government to shirk responsibility and delay action regarding global warming (2010, p. 133, 134 internal footnote omitted).

Antilla's (2005) research, in turn, confirms Boykoff and Boykoff's (2004) finding that balanced coverage does not mean accurate coverage—that journalistic "balance" can equal informational "bias" if the issue that is being discussed is largely consensual within the scientific community. Analyzing the frames constructed by more than 250 US newspapers and wire/news services based in 43 states and the District of Columbia between March 2003 and February 2004, Antilla (2005) found while climate change was not a prime news topic for the newspapers, articles that framed climate change in terms of controversy, debate, or uncertainty were plentiful. According to Antilla, in light of the growing scientific consensus on climate change, the newspapers gave a disproportionate amount of attention to contrarian positions—a finding consistent with that of Boykoff and Boykoff (2004,

p. 126); Jacques et al. (2008, p. 356); McCright and Dunlap (2003, p. 365). Antilla (2005, p. 340) noted that "[i]n order to provide balance while reporting on climate change, some journalists include[d] rebuttals by experts who, often through think-tanks, are affiliated with the fossil fuel industry. Regrettably, this creates the impression that scientific opinion is evenly divided or completely unsettled." Consistent with Boykoff and Boykoff's (2004), Antilla's (2005, p. 350) study demonstrated that "[n]ot only were there many examples of journalistic balance that led to bias, but some of the news outlets repeatedly used climate skeptics—with known fossil fuel industry ties—as primary definers"—which she defined as "those individuals who help frame and define not only what the issues are but also the terms of reference for their discussion" (2005, p. 340 citation and quotation marks omitted). "Worse yet," Antilla (2005, p. 350) continued, "in some instances, such articles originated from wire or news service providers (including newspapers that provide such services or are affiliated with news service agencies)—which caused the exponential spread of misinformation." Finally, Antilla (2005, p. 350) observed that a number of news items were based on scientific research that had not been published in refereed journals, leading her to conclude that media attention to non-peer-reviewed work of contrarians can "perpetuate the myth of a lack of international scientific consensus on anthropogenic climate change—and thereby succeed in maintaining public confusion"—a point alluded to above.

The work of researchers such as Boykoff and Boykoff (2004) and Antilla (2005) reveals that the journalistic norm of balance can compete with the journalistic value of accuracy, leading to biased depictions of knowledge on global climate change in the US prestige press—an unwarranted weight of those that deny its anthropogenic origins or that the problem is even scientifically provable. While "[t]he continuous juggling act journalists engage in, often mitigates against meaningful, accurate, and urgent coverage of the issue of global warming" (Boykoff and Boykoff 2004, p. 125; see also Smith 2005, p. 1474), climate change contrarians also benefit from another dynamic: they do not have to convince anybody of anything. Climate change contrarians simply have to create a sense of doubt. They bear no *burden of production*, to borrow a legal term; they do not have to introduce enough to have a given issue considered by the public or politicians. They bear no *burden of persuasion*, to appropriate another legal term; they do not have to convince anyone to view facts in a way that favors their position. They simply have to instill doubt—and not even *reasonable doubt*, to employ a third legal term. Climate change contrarians "win" through confusion. As McCright (2007, pp. 201–202) explains, "[c]onventional scientists try to validate new knowledge claims about complex climate phenomena, which challenge the dominant social paradigm about how humans interact with the environment. On the other hand, the fossil fuels industry, conservative think tanks, and the contrarians they promote advance their objective of maintaining the status quo merely by obstructing communication of these new knowledge claims. Only a minimal amount of confusion about climate change may be necessary to reinforce the social inertia that perpetuates the status quo, even in the face of considerable scientific evidence otherwise. Thus the goals of the contrarians are achieved more easily than are the goals of conventional climate scientists" (internal citation omitted).

Whereas mainstream climate scientists and environmental groups who support individual and collective efforts to reduce greenhouse gas emissions must convince people (on individual, state, federal, and international levels) *to act*—to change their behaviors or to agree to legislation to moderate or mitigate anthropogenic climate change—climate change contrarians win with *inaction* [see McCright and Dunlap 2000, p. 509, citing Hirschman (1991); McCright and Dunlap (2003, p. 366)]. For climate change contrarians, *inaction = victory*.

As noted above, Lynch et al. (2010) focus on a dynamic of "climate change contrarianism" (although they do not use the term) to argue that the collusion between the Bush Administration and corporate–private interests constitutes a form of state-corporate crime. They make a compelling case, and I would like to build on their argument to offer an interpretation of the broader phenomenon of "climate change contrarianism." Climate change contrarians—the whole "carbon club" (Leggett 2001) or "denial machine" (Begley 2007) term made up of industry lobbies, interest groups, and PR firms—have "maintain[ed] an illusion of intense controversy" (Antilla 2005, p. 340). But the "dueling scientists scenario" (McCright and Dunlap 2000, p. 500; McCright and Dunlap 2003, p. 366; see also Jacques et al. 2008, p. 356) is, indeed, a mirage. While the competition of wide-ranging ideas in free, transparent discourse is integral to liberal democracy, and "[w]hile genuine criticism of various climate science knowledge claims is a valid and common process to advance the science, contrarians demonstrate ulterior questionable motives when they ally themselves consistently with fossil fuels organizations and conservative think tanks to convey their counterclaims outside of the scientific community's normal outlets" (McCright 2007, p. 206). Climate change contrarians do not enhance climate science knowledge nor do they contribute to the "marketplace of ideas" (see generally Brisman 2003). Rather, climate change contrarians have created a "cultural silence," to use Websdale and Ferrell's (1999, pp. 349–350) term.

Websdale and Ferrell employ the "cultural silence" trope to refer to "the socio-historical inattention to phenomena that appear to warrant a deviant label or, indeed, later come to attract such an attribution" (Websdale and Ferrell 1999, pp. 349–350). For example, throughout the 1980s, there were several hundred violent attacks on facilities that provide abortions and other family planning and women's health services. As Jenkins (1999, 2009) explains, the numerous bombings of abortion clinics shared many of the characteristics of other destructive and violent behavior labeled "terrorism," but these bombings were not socially defined as "terrorism" and the word "terrorism" was virtually never used in this context until after the fall 1992 election and change of administration in early 1993. As Jenkins (2009, p. 43) explains, "[i]f agencies refuse to define an issue as grave or threatening, then the media follow suit. The FBI denied for years that abortion-related violence was terrorism, and so it was not classified thus, whether in newspapers, television reports or in the works of academic papers." Continuing Jenkins' line of thought, Websdale and Ferrell (1999, p. 358) observe that "[t]he failure to construct is a potent reminder that it is not the nature of violent or deviant acts per se that ensures legal sanction or, more significantly, a place in the annals of history, but rather the social definition of those acts."

As suggested throughout this chapter, climate change contrarians' ability to undermine the science of climate change and to send and perpetuate mixed or conflicting messages about climate change has resulted in the lack of actual action to address climate change. Indeed, as McCright and Dunlap (2003, p. 352) contend, powerful corporate–political interests have often been successful in keeping environmental problems and technological risks off the political agenda, and in preventing such conditions from becoming widely defined as "problems" in the first place. In the context of global climate change, they argue that "our nation's failure to enact a significant climate policy is heavily influenced by the success of the conservative movement in challenging the legitimacy of global warming as a social problem. Through an abundance of activities, such as flooding the media with brief press releases, holding policy forums, and sponsoring press conferences for policy-makers ... the conservative movement and especially conservative think tanks appear to have successfully affected our nation's policy-making, ... with international implications" (McCright and Dunlap 2003, p. 367, 370). Elsewhere, McCright (2007, p. 209) contends, "[p]olicy-making on science-related issues is lacking at best or seriously flawed at worst when productive input from the scientific community is distorted or rejected on the basis of economic or ideological interests." By perverting journalistic notions of fairness and objectivity so that every mention of anthropogenic climate change is prefaced by "some scientists believe ... " followed by "but skeptics contend ..." (see Boykoff and Boykoff 2004), climate change contrarians have succeeded in discouraging individual changes in behavior in residential sectors (e.g., with respect to lightbulbs, lawn equipment, transportation choices) (see generally Billitteri 2008) and keeping public opinion from demanding mandatory government policies to control emissions (Begley 2007), thereby reducing the likelihood that the federal government will step up with policies addressing greenhouse gas emissions (McCright 2007, p. 209; see also Carvalho 2007, p. 237). Although public confusion over climate change often translates into political inaction and policy gridlock (McCright and Dunlap 2003, p. 366; see also Dietz et al. 2007, p. 186), when energy-climate legislation actually does begin to germinate, climate change contrarians have succeeded in killing those bills (see Begley 2007; McCright and Dunlap 2000, p. 518; see also Friedman 2010a, p. WK9, who asserts that "the totally bogus 'discrediting' of climate science has had serious implications. For starters, it helped scuttle Senate passage of the energy-climate bill needed to scale U.S.-made clean technologies, leaving America at a distinct advantage in the next great global industry").

If anthropogenic climate change constitutes "a crime against the human and non-human inhabitants of the world" (Lynch et al. 2010, p. 215 citing Gelbspan 2004) *and* climate change contrarians' sustained distortion of climate science and persistent assault communication between the climate science community and policy-makers and the general public regarding climate change has served to perpetuate individual and governmental inertia on climate change, then we might consider climate change contrarianism to be a form of "cultural silence." "Unconstructed deviance leaves cultural silences about issues," Websdale and Ferrell (1999, p. 357) explain. In the case of the numerous bombings of abortion clinics throughout the 1980s, the failure to label the bombing campaigns as "terrorism" created a "cultural silence." Similarly, the failure to acknowledge the collusion between the Bush Administration and corporate–private interests as a form of "state-corporate crime," as well as climate change contrarians' advocacy against taking action on climate change, leaves or creates a "cultural silence" about climate change.

Two significant differences bear mention. First, as Jenkins (2009, p. 44) explains, the lack of construction frequently ensures that problems do not remain in public memory. In the instance of climate change, it is the lack of construction *combined with* the partisan and polarized debate (generated by climate change contrarians' messages about scientific uncertainty) that I have argued may be contributing to the mainstream American public's "climate fatigue" (Kerr 2009; see also Dietz et al. 2007, p. 210; Dispensa and Brulle 2003; Tákacs-Sánta 2007, pp. 31–32)—a situation in which all the talk about climate change has produced little in the way of actual action (which is exactly what climate change contrarians want). Thus, whereas Websdale and Ferrell, as well as Jenkins, contemplate "cultural silence" as sociohistorical inattention to a phenomenon that warrants a deviant label, in the case of climate change, we are witnessing how unwarranted overanalysis of a phenomenon has resulted in *analysis paralysis*. The mainstream American public has gotten turned off—not just from the "debate" about climate change, but from the issue and its urgency itself [see Moser and Dilling (2004, p. 37)].

Second, the "cultural silence" created by climate change contrarianism has implications that extend far beyond climate change. Lynch et al. (2010, p. 230) claim that the collusion between the administration of George W. Bush and corporations not only undermined the science of climate change and climate change policy—and not only contributed to harmful human/public/environmental health consequences—but that it has had a negative "effect on public perceptions of government and public cynicism concerning government and the legitimacy of state institutions." Climate change contrarianism runs the risk of rendering science as "dismissable endeavor," to borrow Carvalho's (2007, p. 238) phrase.

Scientific ideas are inherently provisional and capable of being overturned when better answers are discovered through evidence generated from experimentation and observation (see, e.g., Gleick et al. 2010; Smith 2005, p. 1474). While scientific progress is propelled by curiosity, it is "hindered by a cottage industry of engagement with 'climate skeptics,' few of which have scientific credentials or a true interest in the scientific process" (Moser and Dilling 2004, p. 38box). Indeed, climate change contrarians have made a mockery of the scientific process and placed the very definition of "science" at stake, transforming the contextual and contingent nature of science so that it now appears "plural and open-ended" (Carvalho 2007, p. 238). This abuse of science to carry out what is really a debate or conflict about *values* has undermined the integrity of science and, indeed, the entire scientific enterprise (see, e.g., McCright 2007, p. 207)—which could result in or create other "cultural silences."

Conclusion

Why is it important to look at media representations of climate change? What bearing does such an analysis have on conservation, ecoglobal, or green criminologists, in general, or those specifically working on climate change?

As discussed above, the media does not just provide information for the public, nor does it merely *report* on various events. Rather, it *presents* information and events, *represents* them, and transforms them into *issues*. In so doing, the media transmit emotion, which affects how the public forms its opinions about issues, which subsequently affects individual and collective decision-making and behavior, as well as establish public priorities.

In the context of environmental matters, while it is true that "[i]ssues pertaining to the protection of the planet continue to capture media headlines and continually focus public and political debate" (Walters 2007, p. 198), it is also the case that the mainstream press underrepresents environmental risks and individual instances of environmental crime and harm, as well as the seriousness of such crimes and harms. In the specific context of global climate change, this chapter has argued that corporate–political entities have taken advantage of the media's duty to report both sides, meaning that climate change contrarianism has received undue attention. This "balance," however, equals "bias" when the issue "is largely consensual within the scientific community" (Carvalho 2007, p. 233), and this chapter has contended that such bias has contributed to individual and collective reluctance to address global climate change.

Cultural criminologists have examined the nature of and meanings conveyed by mediated images of street-level crime, deviance, and violence, and have documented the history of political–corporate attempts to augment fear and depict a problem as *greater* than it may be when such *embellishment* is expedient or otherwise beneficial (i.e., "moral panics" and "folk devils"). Green criminologists may wish to follow cultural criminologists' lead and explore the nature of and ways in which the media present information about environmental crimes, harms and risks, whether and how the media construct environmental issues, whether and how the media transmit emotion about environmental crimes, harms, issues and risks, and how the public responds—how the public forms its mind, establishes its priorities, and chooses (or chooses not) to act (individually or collectively). Green criminologists may wish to ask: What environmental crimes, harms, and risks are being reported? What environmental crimes, harms, and risks are *not* being reported? What causes and consequences (if any) are being reported? Who bears the responsibility for environmental crimes, harms, and risks. At the same time, green criminologists may find a need to engage in a somewhat different process than cultural criminological endeavors—specifically, to situate current climate change contrarianism within a history of corporate–political *defiance, refutation, and rejectio*n of phenomena that hurt their interests.

In public surveys, climate change frequently does not fall within the top ten priority issues when compared with the economy, education, employment, crime, and other such concerns (see, e.g., Moser and Dilling 2004, p. 36, 37). Green criminologists can play a vital role in helping to reorient anthropocentric climate change *as* crime. White (2009, p. 6) contends that one duty of green criminology is to name environmental harms as "'criminal', even if not considered illegal in conventional terms. Those who determine and shape the law are very often those whose activities need to be criminalized for the sake of planetary well-being." If various activities are creating "environmental harm"—and if various corporate–political entities are attempting to downplay or outright dispute this harm—then we need to (continue to) examine and challenge the forces that are working *against this*—namely, the corporate–political powers that have fostered a "cultural silence" of and around climate change via contrarian messages delivered through the media.

References

Adam, D. (2006, September 20). Royal Society tells Exxon: stop funding climate change denial. *The Guardian*. http://www.guardian.co.uk/environment/2006/sep/20/oilandpetrol.business.

Adam, D. (2010, February 20). Climate scientist says Himalayan glacier report is 'robust and rigorous'. *The Guardian*. http://www.guardian.co.uk/environment/2010/feb/14/climate-scientist-himalayan-glacier-report.

Anderegg, W. R. L. (2010). Diagnosis earth: The climate change debate. *Thought & Action: The NEA Higher Education Journal, 25*, 23–36.

Antilla, L. (2005). Climate of scepticism: US newspaper coverage of the science of climate change. *Global Environmental Change, 15*, 338–352.

Associated Press. (2010, August 31). Virginia case against climate researcher is rejected. *The New York Times*, A14.

Bagla, P. (2009, December 5). Himalaya glaciers melting deadline 'a mistake.' *BBC News*. http://news.bbc.co.uk/2/hi/8387737.stm.

Bagla, P. (2010). Newmaker interview: Climate science leader Rajendra Pachauri confronts the critics. *Science, 327*, 510–511.

Barnett, A., & Townsend, M. (2004, November 28). Greenhouse effect 'may benefit man': Claims by think-tank outrage eco-groups. *The Guardian*. http://www.guardian.co.uk/society/2004/nov/28/environment.environment.

Barr, S. (1997, July 2). 12 Foundations Push Conservative Agenda; $210 Million Given Over 3 Years. *The Washington Post*. 21, 4.

Barringer, F. (2011, 28 July). Cataclysms, climate change and California opinion. *The New York Times: Green: A Blog About Energy and the Environment*. http://green.blogs.nytimes.com/2011/07/28/cataclysms-climate-change-and-california-opinion/?scp=1&sq=climategate&st=cse.

Becker, B., & Hulse, C. (2010, June 17). Barton: what I really mean to say. *The New York Times*.

Beder, S. (1999). Corporate hijacking of the greenhouse debate. *The Ecologist, 29*(2), 119–122.

Begley, S. (2007, August 13). The truth about denial. *Newsweek*. http://www.thedailybeast.com/newsweek/2007/08/13/the-truth-about-denial.html.

Begley, S. (2010, June 25). Newspapers retract 'Climategate' claims, but damage still done. *Newsweek*. http://www.thedailybeast.com/newsweek/blogs/the-gaggle/2010/06/25/newspapers-retract-climategate-claims-but-damage-still-done.html.

Benton, T. (2007). Ecology, community and justice: The meaning of green. In B. Piers & S. Nigel (Eds.), *Issues in green criminology: Confronting harms against environments, humanity and other animals* (pp. 3–31). Cullompton, Devon: Willan Publishing.

Billitteri, T. J. (2008). Reducing your carbon footprint: Can individual actions reduce global warming? *CQ Researcher, 18*(42), 985–1008.
Boekhout van Solinge, T. (2010). Deforestation crimes and conflicts in the Amazon. *Critical Criminology, 18*(4), 263–77.
Boykoff, M. T., & Boykoff, J. M. (2004). Balance as bias: Global warming and the US Prestige Press. *Global Environmental Change, 14*, 125–36.
Brisman, A. (2003). Rethinking the Case of Matthew F. Hale: fear and loathing on the part of the Illinois Bar Committee on Character and Fitness. *Connecticut Law Review, 35*(3), 1399–1424.
Brisman, A. (2005). The aesthetics of wind energy systems. *New York University Environmental Law Journal, 13*(1), 1–133.
Brisman, A. (2008). Crime-environment relationships and environmental justice. *Seattle Journal for Social Justice, 6*(2), 727–817.
Brisman, A. (2009, October 7). *Resisting Speed.* Paper presented at First Annual International Crime, Media & Popular Culture Studies Conference: A Cross Disciplinary Exploration, Indiana State University Terre Haute, Indiana.
Brisman, A. (2010a). 'Creative crime' and the phytological analogy. *Crime Media Culture, 6*(2), 205–225.
Brisman, A. (2010b). The indiscriminate criminalisation of environmentally beneficial activities. In W. Rob (Ed.), *Global environmental harm: Criminological perspectives* (pp. 161–92). Devon, UK: Willan Publishing.
Brisman, A. (2010/2011). Fictionalized criminal law and youth legal consciousness. *New York Law School Law Review, 55*(4), 1039–1071.
Brisman, A., & South, N. (2010, November 17). *A Green Agenda for Cultural Criminology*. Paper presented at the Annual Meeting of the American Society of Criminology, San Francisco, CA.
Broder, J. M. (2010, October 21). Skepticism on climate change is article of faith for tea party. *The New York Times*, A1, A4.
Calmes, J. (2010a). *White House throws the shame word back at Barton*. June: The New York Times. 17.
Calmes, J. (2010b, June 18). Republican lawmaker, under pressure, backpedals from apology to BP. *The New York Times*.
Carter, N. (2007). *The politics of the environment: Ideas, activism, policy* (2nd ed.). Cambridge: Cambridge University Press.
Carvalho, A. (2005). Representing the politics of the greenhouse effect: Discursive strategies in the British media. *Critical Discourse Studies, 2*(1), 1–29.
Carvalho, A. (2007). Ideological cultures and media discourses on scientific knowledge: Re-reading news on climate change. *Public Understanding of Science, 16*, 223–243.
Castells, M. (2004). *The power of identity: The information age: Economy, society, and culture* (2nd ed.). Oxford: Blackwell.
Corcoran, T. (2010, January 6). The cool down in climate polls. *National Post*. http://www.nationalpost.com/opinion/columnists/story.html?id=e06924ca-91e0-4a18-8165-126656414605&p=1.
Crowley, M. (2006, May 22). Barton Fink. *The New Republic*, 8–10.
Denton, F. (2000). Gendered impacts of climate change—A human security dimension. *Energia News, 3*(3), 13–14.
Denton, F. (2002). Climate change vulnerability, impacts, and adaptation: Why does gender matter? *Gender and Development, 10*(2), 10–20.
Dietz, T., Dan, A., & Shwom, R. (2007). Support for climate change policy: Social psychological and social structural influences. *Rural Sociology, 72*(2), 185–214.
Dispensa, J. M., & Brulle, R. J. (2003). Media's social construction of environmental issues: Focus on global warming–a comparative study. *International Journal of Sociology and Social Policy, 23*, 74–105.
Douthat, R. (2010, July 26). The right and the climate. *The New York Times*, A23.

Doyle, J. (2007). Picturing the clima(c)tic: Greenpeace and the representational politics of climate change communication. *Science as Culture, 16*(2), 129–50.
Easterbrook, G. (2011, April 24). State of Petroshock. *The New York Times Book Review*, 12.
Editorial. (2005, July 23). Hunting witches. *The Washington Post*. http://www.washingtonpost.com/wp-dyn/content/article/2005/07/22/AR2005072201658.html.
Editorial. (2008, January 1). The one environmental issue. *The New York Times*, A18.
Editorial. (2009, December 6). That climate change e-mail. *The New York Times*, WK9.
Editorial. (2010a). Science subpoenaed. *Nature, 465*(7295), 135–36.
Editorial. (2010b, May 24). Are they paying attention? *The New York Times*, A24.
Editorial. (2010c, July 10). A climate change corrective. *The New York Times*.
Editorial. (2010d, September 21). The brothers Koch and AB32. *The New York Times*, A30.
Editorial. (2010e, October 18). In climate denial, again. *The New York Times*, A34.
Eilperin, J. (2008, November 19). Obama addresses climate summit. *The Washington Post*. http://www.washingtonpost.com/wp-dyn/content/article/2008/11/18/AR2008111803286.html.
Ferrell, J. (1995). Culture, crime, and cultural criminology. *Journal of Criminal Justice and Popular Culture, 3*(2), 25–42. http://www.albany.edu/scj/jcjpc/vol3is2/culture.html.
Ferrell, J. (1999). Cultural Criminology. *Annual Review of Sociology, 25*, 395–418.
Ferrell, J. (2003). Cultural criminology. In M. D. Schwartz & S. E. Hatty (Eds.), *Controversies in critical criminology* (pp. 71–84). Cincinnati, OH: Anderson Publishing Company.
Ferrell, J. (2004). Boredom, crime and criminology. *Theoretical Criminology, 8*(3), 287–302.
Ferrell, J. (2006). *Empire of scrounge: Inside the urban underground of dumpster diving, trash picking, and street scavenging*. New York: New York University Press.
Ferrell, J. (2009). Crime and culture. In C. Hale, K. Hayward, K. Wahidin, & E. Wincup (Eds.), *Criminology* (2nd ed., pp. 157–75). Oxford: Oxford University Press.
Ferrell, J., Hayward, K., & Young, J. (2008). *Cultural criminology: an invitation*. London: Sage.
Ferrell, J., & Sanders, C. R. (1995). Toward a cultural criminology. In J. Ferrell & C. R. Sanders (Eds.), *Cultural criminology* (pp. 297–326). Boston: Northeastern University Press.
Ferrell, J., & Websdale, N. (1999). Materials for making trouble. In J. Ferrell & N. Websdale (Eds.), *Making trouble: cultural constructions of crime, deviance, and control* (pp. 3–21). Hawthorne, NY: Aldine de Gruyter.
Fitzgerald, A., & Baralt, L. B. (2010). media constructions of responsibility for the production and mitigation of environmental harms: The case of mercury-contaminated fish. *Canadian Journal of Criminology and Criminal Justice, 52*(4), 341–368.
Friedman, T. L. (2010a, September 19). Aren't we clever. *The New York Times*, WK9.
Friedman, T. L. (2010b, February 17). Global weirding is here. *The New York Times*, A23.
Friedman, T. L. (2010c, July 25). We're gonna be sorry. *The New York Times*, WK8.
Gelbspan, R. (1998). *The heat is on: The climate crisis; the cover-up; the prescription*. Cambridge: MAL Perseus Books.
Gelbspan, R. (2004). *Boiling point: How politicians, big oil and coal, journalists, and activists have fueled the climate crisis—and what we can do to avert disaster*. New York: Basic Books.
Gelbspan, R. (2005). Snowed. *Mother Jones, June*, 42–43.
Gillis, J. (2010a, July 2). Penn State Panel clears scientist over 'Climategate'. *The New York Times*, A14.
Gillis, J. (2010b, July 8). Panel, in report, clears scientists of rigging climate change data. *The New York Times*, A9.
Gillis, J. (2010c, August 15). In weather chaos, a case for global warming. *The New York Times*, 1, 12.
Gleick, P. H., Adams, R. M., Amasino, R. M., Anders, E., Anderson, D. J., Anderson, W. W., et al. (2010). Letter to the Editor: Climate change and the integrity of science. *Science, 328*, 689–690.
Gore, A. (2006). *An inconvenient truth*. New York: Melcher Media.
Greer, C. (2009). Crime and media: Understanding the connections. In C. Hale, K. Hayward, A. Wahidin, & E. Wincup (Eds.), *Criminology* (2nd ed., pp. 177–203). Oxford: Oxford University Press.
Hansen, A. (1991). The media and the social construction of the environment. *Media, Culture and Society, 13*, 443–458.

Hayward, K. J., & Morrison, W. (2009). Theoretical criminology: A starting point. In C. Hale, K. Hayward, A. Wahidin, & E. Wincup (Eds.), *Criminology* (2nd ed., pp. 73–102). Oxford: Oxford University Press.

Hayward, K. J., & Young, J. (2004). Cultural criminology: Some notes on the script. *Theoretical Criminology, 8*(3), 250–273.

Helderman, R. S. (2010, August 31). Judge rejects Ken Cuccinelli's probe of U-Va global warming records. *The Washington Post,* B1.

Hirschman, A. (1991). *The rhetoric of reaction*. Cambridge: Harvard University Press.

Hulme, M. (2009). *Why we disagree about climate change: Understanding controversy, inaction and opportunity*. Cambridge: Cambridge University Press.

IPCC. (2010, January 20). IPCC statement on the melting of Himalayan glaciers. http://www.ipcc.ch/pdf/presentations/himalaya-statement-20january2010.pdf.

Jacques, P. J., Dunlap, R. E., & Freeman, M. (2008). The organization of denial: Conservative think tanks and environmental scepticism. *Environmental Politics, 17*(3), 349–385.

Jancell, M. L. (2007). *Environmental crime and the media: News coverage of petroleum refining industry violations*. New York: LFB Scholarly.

Jenkins, P. (1999). Fighting terrorism as if women mattered: Anti-abortion violence as unconstructed terrorism. In F. Jeff & N. Websdale (Eds.), *Making trouble: Cultural constructions of crime, deviance, and control* (pp. 319–46). New York: Aldine de Gruyter.

Jenkins, P. (2009). Failure to launch: Why do some social issues fail to detonate moral panics? *British Journal of Criminology, 49*(1), 35–47.

Kakutani, M. (2010, August 3). Warming is real. Now what? *The New York Times*, C1, C4.

Kaufman, L. (2010a, March 4). Darwin foes add warming to target list. *The New York Times*, A1, A4.

Kaufman, L. (2010b, October 19). Kansans Scoff at global warming but embrace cleaner energy. *The New York Times*, A1, A4.

Kaufman, L. (2011, May 13). Scientists' report stresses urgency of limiting greenhouse gas emissions. *The New York Times*, A16.

Kerr, R. A. (2009). Amid worrisome signs of warming, 'Climate fatigue' sets in. *Science, 326*, 926–928.

Kintisch, E. (2009). Stolen E-mails turn up heat on climate change rhetoric. *Science, 326*, 1329.

Krugman, P. (2010a, May 3). Drilling, disaster, denial. *The New York Times*, A25.

Krugman, P. (2010b, July 26). Who cooked the planet? *The New York Times*, A23.

Leggett, J. (2001). *The carbon war: Global warming and the end of the oil era.* New York: Routledge.

Lizza, R. (2010, October 11). As the world burns. *The New Yorker*, 70–77, 79–83.

Lowe, T., Brown, K., Dessai, S., de França, M., Doria, K. H., & Vincent, K. (2006). Does tomorrow ever come? Disaster narrative and public perceptions of climate change. *Public Understanding of Science, 15*, 435–457.

Lynch, M. J., Burns, R. G., & Stretesky, P. B. (2010). Global warming and state-corporate crime: The politicization of global warming under the Bush administration. *Crime Law and Social Change, 54*(3–4), 213–239.

Lynch, M. J., Nalla, M., & Miller, K. (1989). Cross-cultural perceptions of deviance: The case of Bhopal. *Journal of Research in Crime and Delinquency, 26*(1), 7–35.

Lynch, M. J., & Stretesky, P. B. (2010). Global warming, global crime: a green criminological perspective. In R. White (Ed.), *Global environmental harm: Criminological perspectives* (pp. 62–84). Devon, UK: Willan Publishing.

Lynch, M. J., Stretesky, P., & Hammond, P. (2000). Media coverage of chemical crimes Hillsborough County, Florida, 1987–97. *British Journal of Criminology, 40*, 112–126.

Marx, E. (2008, November 10). Are you reading to go on a carbon diet? *The Christian Science Monitor*. http://www.csmonitor.com/Business/2008/1110/p13s01-wmgn.html.

McCarthy, M., Brennan, M., De Boer, M., & Ritson, C. (2008). Media risk communication: What was said by whom and how was it interpreted. *Journal of Risk Research, 11*(3), 375–394.

McCright, A. M. (2007). Dealing with climate change contrarians. In S. C. Moser & L. Dilling (Eds.), *Creating a climate for change: Communicating climate change and facilitating social change* (pp. 200–12). Cambridge: Cambridge University Press.

McCright, A. M., & Dunlap, R. E. (2000). Challenging global warming as a social problem: An analysis of the conservative movement's counter-claims. *Social Problems, 47*(4), 499–522.

McCright, A. M., & Dunlap R. E. (2003). Defeating Kyoto: The Conservative Movement's impact on U.S. climate change policy. *Social Problems, 50*(3), 348–373.

McKay, G. (1998). DiY culture: Notes towards an intro. In G. McKay (Ed.), *DiY culture: Party & protest in nineties Britain* (pp. 1–53). London: Verso.

Moser, S. C., & Dilling, L. (2004). Making climate hot: Communicating the urgency and challenge of global climate change. *Environment, 46*(10), 32–46.

National Research Council, Division on Earth and Life Studies, Board on Atmospheric Sciences and Climate, & Committee on America's Climate Choices. (2011). *America's Climate Choices.* Washington, D.C.: The National Academies Press. http://dels.nas.edu/Report/Americas-Climate-Choices/12781.

Nelkin, D. (1995). *Selling science: How the press covers science and technology*. New York: W.H. Freeman. Revised edition.

Newport, F. (2010, March 11). Americans' global warming concerns continue to drop. Washington, D.C.: Gallup, Inc. http://www.gallup.com/poll/126560/Americans-Global-Warming-Concerns-Continue-Drop.aspx.

Nissani, M. (1999). Media coverage of the greenhouse effect. *Population and Environment: A Journal of Interdisciplinary Studies, 21*(1), 27–43.

Ockwell, D., Whitmarsh, L., & O'Neill, S. (2009). Reorienting climate change communication for effective mitigation: Forcing people to be green or fostering grass-roots engagement? *Science Communication, 30*(3), 305–327.

Patterson, M. (2010, September 2). Meltdown of the climate 'consensus.' *The New York Post.* http://www.nypost.com/p/news/opinion/opedcolumnists/meltdown_of_the_climate_consensus_G0kWdclUvwhVr6DYH6A4uJ.

Pollack, H. N. (2003). *Uncertain science .. uncertain world.* Cambridge: Cambridge University Press.

Revkin, A. C. (2009a, November 21). Hacked e-mail is new fodder for climate dispute. *The New York Times.*

Revkin, A. C. (2009b, November 28). Hacked E-mail data prompts calls for changes in climate research. *The New York Times*, A8.

Rosenthal, E. (2010a, May 25). Climate fears turn to doubts among Britons. *The New York Times*, A1, A8.

Rosenthal, E. (2010b, January 19). U.N. Panel's glacier warning is criticized as exaggerated. *The New York Times*, A8.

Ruggiero, V., & South, N. (2010). Critical criminology and crimes against the environment. *Critical Criminology, 18*(4), 245–250.

Safina, C. (2011). *A sea in flames: The deepwater horizon oil blowout.* New York, NY: Crown.

Scarce, R. (1990). *Eco-warriors: Understanding the radical environmental movement.* Chicago: The Noble Press.

Schoenfeld, A. C., Meier, R. F., & Griffin, R. J. (1979). Constructing a social problem: The press and the environment. *Social Problems, 27*(1), 38–61.

Schulman, S., et al. (2007). *Smoke, mirrors & hot air: How ExxonMobil uses big tobacco's tactics to manufacture uncertainty on climate change.* Cambridge, MA: Union of Concerned Scientists. http://www.ucsusa.org/assets/documents/global_warming/exxon_report.pdf.

Simon, D. R. (2000). Corporate environmental crimes and social inequality. *American Behavioural Scientist, 43*, 633–345.

Smith, J. (2005). Dangerous news: Media decision making about climate change risk. *Risk Analysis, 25*(6), 1471–1482.

Spector, M., & Kitsuse, J. I. (1977). *Constructing social problems.* Menlo Park, CA: Cummings.

Stevens, W. K. (2007, February 6). On the climate change beat, doubt gives way to certainty. *The New York Times*, D1, D3.
Sussman, B. (2010). *Climategate: A veteran meteorologist exposes the global warming scam*. Washington, D.C.: WND Books.
Tákacs-Sánta, A. (2007). Barriers to environmental concern. *Human Ecology Review, 14*(1), 26–38.
The New York Post. (2010). The EPA's climate con. 22 Feb. http://www.nypost.com/p/news/opinion/editorials/the_epa_climate_con_0V5iFmpmnuneEYGxvnI22J.
Tierney, J. (2009, December 1). E-mail fracas shows peril of trying to spin science. *The New York Times*, D1, D2.
Turner, T. (1993). The role of indigenous peoples in the environmental crisis: The example of the Kayapo of the Brazilian Amazon. *Perspectives in Biology and Medicine, 36*(3), 526–545.
United Nations Environment Programme. (2009). *Climate Change Science Compendium 2009*. http://www.unep.org/pdf/ccScienceCompendium2009/cc_ScienceCompendium2009_full_highres_en.pdf.
Wachholz, S. (2007). 'At risk': Climate change and its bearing on women's vulnerability to male violence. In B. Piers & N. South (Eds.), *Issues in green criminology: Confronting harms against environments, humanity and other animals* (pp. 161–85). Cullumpton, Devon: Willan Publishing.
Walters, R. (2007). Crime, regulation and radioactive waste in the United Kingdom. In B. Piers & N. South (Eds.), *Issues in green criminology: Confronting harms against environments, humanity and other animals* (pp. 186–205). Cullompton, Devon: Willan Publishing.
Warner, J. (2011, February 27). Fact-free science. *The New York Times*, 11–12.
Weart, S. R. (2003). *The discovery of global warming*. Cambridge, MA: Harvard University Press.
Websdale, N., & Ferrell, J. (1999). Taking the trouble: Concluding remarks and future directions. In F. Jeff & N. Websdale (Eds.), *Making trouble: Cultural constructions of crime, deviance, and control* (pp. 349–64). Hawthorne, NY: Aldine de Gruyter.
Webster, B. (2010, February 15). UN must investigate warming 'bias', says former climate chief. *London Times*. http://www.timesonline.co.uk/tol/news/environment/article7026932.ece and http://www.timesonline.co.uk/tol/news/environment/article7026932.ece#cid=OTC-RSS&attr=797093.
White, R. (2007). Green criminology and the pursuit of social and ecological justice. In B. Piers & N. South (Eds.), *Issues in green criminology: Confronting harms against environments, humanity and other animals* (pp. 32–54). Cullompton, Devon: Willan Publishing.
White, R. (2008). *Crimes against nature: Environmental criminology and ecological justice*. Cullumpton, Devon: Willan Publishing.
White, R. (2009). Globalisation and environmental harm. In R. White (Ed.), *Global environmental harm: Criminological perspectives* (pp. 3–19). Cullompton, Devon: Willan Publishing.
Whitmarsh, L. (2009). What's in a name? Commonalities and differences in public understanding of "climate change" and "global warming". *Public Understanding of Science, 18*, 401–420.
Whyte, W. F. (1993) [1943]. *Street Corner Society: The social structure of an Italian slum* (4th ed.). Chicago: University of Chicago Press.
Wilson, K. M. (1995). Mass media as sources of global warming knowledge. *Mass Communication Review, 22*(1–2), 75–89.
Wilson, K. M. (2000a). Drought, debate, and uncertainty: Measuring reporters' knowledge and ignorance about climate change. *Public Understanding of Science, 9*, 1–13.
Wilson, K. M. (2000b). Communicating climate change through the media: Predictions, politics and perceptions of risk. In A. Stuart, A. Barbara, & C. Cynthia (Eds.), *Environmental risks and the media* (pp. 201–17). London: Routledge.
Yates, R., Powell, C., & Beirne, P. (2001). Horse maiming in the English countryside: Moral panic, human deviance, and the social construction of victimhood. *Society and Animals, 9*, 1–23.

Chapter 5
Is Global Warming a State-Corporate Crime?

Ronald C. Kramer and Raymond J. Michalowski

The term "crime" has become increasingly common in critical commentary concerning global warming and the associated environmental harm resulting from climate change. In 2009, for instance, Hans Joachim Schellnhuber, a physicist and the chief climate advisor to the government of Germany referred to 8 years of inaction on global warming and climate change by the George W. Bush administration in the United States by saying, "This was a crime" (quoted in Hertsgaard 2011, p. 254). On the academic side, criminologists Lynch et al. (2010, p. 215) similarly condemn the Bush administration for criminal and "callous neglect of the issue of global warming." Contending that global warming has "criminological and sociological relevance on several levels," they explore "the politicalization of global warming under the Bush administration" as an example of *state-corporate crime* (Lynch et al. 2010, p. 213).

This chapter expands on this notion of global warming and climate change as state-corporate crime by examining how transnational corporations, particularly in the fossil fuel industry, and the nation states of the global North, particularly the United States, act in concert in ways that, intentionally or not, cause widespread environmental and social harm. Corporate and state actors in interaction with each other create these harms in four ways (1) by denying that global warming is caused by human activity, (2) by blocking efforts to mitigate greenhouse gas emissions, (3) by excluding progressive, ecologically just adaptations to climate change from the political arena, and (4) by responding to the social conflicts that arise from climate change by transforming themselves into "fortress societies while the rest of

R.C. Kramer (✉)
Department of Sociology, Western Michigan University, 1903 West Michigan Avenue,
Kalamazoo, MI 49008-5257, USA
e-mail: ronald.kramer@wmich.edu

R.J. Michalowski
Department of Criminal Justice, Northern Arizona University, P.O. Box 15005,
Flagstaff, AZ 80611-5005, USA
e-mail: raymond.michalowski@nau.edu

R. White (ed.), Climate Change from a Criminological Perspective,
DOI 10.1007/978-1-4614-3640-9_5, © Springer Science+Business Media New York 2012

the world slips into collapse" (Parenti 2011, p. 20). After examining each of these offenses, we conclude by arguing that if those who consider themselves "green criminologists" hope to contribute to reversing, or at least slowing, global climate change, then they need to engage in a "public criminology" that communicates the relationship between state-corporate crimes and environmental degradation to audiences beyond their academic peers.

Global Warming and Green Criminology

Although concerns about the possible harms associated with anthropogenic global warming have existed since at least the 1970s, Lynch and Stretesky (2010, p. 62) point out that: "Criminologists have been slow to consider climate change as a relevant issue." More recently, however, the development of what some term a "green criminological perspective" (Lynch and Stretesky 2003, 2010; Beirne and South 2006, 2007), and others an "eco-global criminology" (White 2008, 2010, 2011), has led a growing number of criminologists to begin considering the criminological import of global warming. This requires blending criminological insights with existing environmental science. This science has provided clear evidence that state and corporate actors have brought the globe to the brink of environmental collapse, and in doing so have committed and are committing grave *crimes* against, not just humanity, but all life as we know it. Not only are these crimes grave, their threat is immediate. As James Hansen (2009, p. ix), director of the NASA Goddard Space Studies Center, stated: "The startling conclusion is that continued exploitation of all fossil fuels on Earth threatens not only the other millions of species on the planet but also the survival of humanity itself-and the timetable is shorter than we thought."

Currently, there is no established body of international or domestic law that offers a legal framework for criminologists concerned with global warming and the environmental harms flowing from it. There are some efforts underway to change this situation (see Chap. 6). The British lawyer Polly Higgins, for instance, has proposed that *ecocide* be recognized by the United Nations as an international crime along with genocide, crimes against humanity, war crimes, and crimes of aggression. Higgins (2010, p. 63) defines ecocide as: "The extensive destruction, damage to or loss of ecosystem(s) of a given territory, whether by human agency or by other causes, to such an extent that peaceful enjoyment by the inhabitants of that territory has been severely diminished." If adopted, "ecocide" would constitute a legal definition under which global warming and climate change would become, not just an environmental problem, but also an international crime. The likelihood that the United Nations, dominated as it is by the carbon dependent and carbon profiting nations of the global North, would establish ecocide as an international crime is slim. Another effort to bring harmful greenhouse gas emissions within a legal framework is the ongoing lawsuit brought by a number of US states against the five largest American utilities in an effort to have their emissions defined as a "public nuisance" (BusinessGreen 2011a). If successful, this suit could force the federal government to impose more stringent regulations on the fossil fuel industry (BusinessGreen 2011b).

Despite these legal efforts, the emission of carbon dioxide and other greenhouse gases, which accumulate in the Earth's atmosphere where they trap heat that generates both planetary warming and climate change, are currently not defined as crimes under any law. With regard to global warming there is still nothing comparable to the Montreal Protocol, which was adopted in 1987 to eliminate aerosols and other chemicals that were responsible for a growing hole in the Earth's protective ozone layer. This Protocol has been successful in eliminating the hydrofluorocarbons (HFCs) that damage the ozone layer. Consequently, some policy experts have suggested that we should curb global warming by including greenhouse gases under this existing, and demonstrably successful treaty (Broder 2010).

To date, the Kyoto Protocol of 1997 is the only major international accord aimed at regulating and reducing greenhouse gas emissions. Unfortunately, this treaty, "an unambitious little thing" as Dyer (2010, p. 142) calls it, is fatally flawed and has done little to reduce carbon emissions. In the first place, the reductions in greenhouse gases mandated by the treaty were too small and applied to only a few countries. Secondly, climate scientists failed to anticipate that extreme weather events and climatic changes due to global warming would increase as dramatically or quickly as they have. Third, most governments resisted calls for larger reductions, and fast-emerging nations such as China and India won agreements that did not oblige them to curb their emissions at all. The biggest failure of the Kyoto Protocol, however, was that the largest emitter of greenhouse gases, the United States, failed to ratify the treaty. As NASA scientist James Hansen points out: "The U.S. sabotaged the effectiveness of the Kyoto Protocol by not signing on. Without the biggest polluter by far, and without the biggest economy, Kyoto could not be very effective" (quoted in Dyer 2010, p. 143). Thus, in its present form the Kyoto Protocol provides little in the way of a legal framework that criminologists can use as a juridical warrant to address anthropogenic global warming as a state-corporate crime.

The lack of a legal framework on which criminologists can base their study of the emission of greenhouse gases as crimes, however, should not be seen as a barrier to this inquiry (see also Chap. 12). Indeed, unless criminologists escape the juridical trap that mandates they only study that which states, through their law-making systems, tell them is a crime, the social injuries caused by the most powerful actors in the contemporary world, transnational corporations and national states, will remain forever outside their reach (Michalowski 2010). Instead, criminologists concerned with global warming can and should utilize concepts such as *social injury/harm* (Michalowski 1985; Tift and Sullivan 2001; Hillyard et al. 2004; Pemberton 2007) and *state crimes of omission* (Barak 1991; Kauzlarich, Mullins and Matthews 2003) as the starting point for their inquiries. As White (2011, p. 21) points out: " A basic premise of green criminology is that we need to take environmental harm seriously, and in order to do this we need a conceptualization of harm that goes beyond conventional understandings of crime." Lynch and Stretesky (2010, pp. 70–71) argue that green criminology should study the "harms that directly damage the ecosystem or its parts (direct victimization of the environment), or victimize species through ecosystem damage (indirect victimization)." They also note that, insofar as "scientists are continually discovering new ways in which global warming produces harm ... it is important for green criminologists to stay abreast of this literature in order to address the varieties of victimization and emerging crimes and harms science identifies." The mandate is both clear and challenging. Green criminologists must understand the scientific literature about, as well as the sociological implications of, global warming and climate change.

Recent overviews of the scientific research on global warming and climate change demonstrate the catastrophic nature of the harms that are being inflicted on the ecosystem (Dyer 2010; Hamilton 2010; Hansen 2009; McKibben 2010), harms that are more severe and occurring at a faster pace than predicted in the 2007 Intergovernmental Panel on Climate Change (IPCC) report. The burning of fossil fuels has already raised the temperature of the planet by almost 1°C (1.8°F) over the pre-industrial average and some scientists estimate it could go as high as an alarming 5 or 6°C, or 9 to 11°F (Dyer 2010; Hamilton 2010). The concentration of carbon dioxide in the atmosphere has increased from 275 parts per million at the dawn of the industrial age to close to 400 parts per million currently, and it appears to be headed for 550 or 650. James Hansen (2009) argues that the only safe level, that is, one that would not risk global warming, is 350 parts per million. Unfortunately, even if we could get back to that level in the near future, great damage has already been done. Again, the research provides clear evidence of this damage: the melting of the polar ice caps, the Greenland ice sheet and the Himalayan glaciers are already contributing to a rise in sea levels; extreme weather events such as an increase in global rainfall in some areas with intractable droughts in others, more severe heat waves, and more frequent and stronger hurricanes; increasing deforestation, particularly in the Amazon rain forest, and the expansion of the tropics which pushes dry subtropics further ahead; and the acidification of the oceans with corrosive effects on shellfish and coral reefs (Hansen 2009; McKibben 2010).

Hansen (2009) also points out that as global warming continues, positive feedback effects will occur that will in turn trigger runaway heating that is essentially uncontrollable and irreversible for thousands of years. Positive feedback tipping points that will dramatically accelerate the heating of the planet include the loss of the Arctic albedo effect (the loss of reflective ice leads to more dark open water in the Arctic which absorbs more solar radiation), the release of huge quantities of methane from the melting permafrost, and the die-back of the Amazon rainforest. Loss of rainforest, in turn will further increase the concentration of carbon dioxide in the global atmosphere, anywhere from 20 to 200 parts per million by the end of the century, with devastating and potentially lethal impacts on many forms of life (Richardson et al. 2011, p. 86).

A key impediment to an effective response to global warming is that the consequences of greenhouse gas emissions involve complex causal chains. As Hamilton (2010, p. 25) points out: "The lag between emissions and their effects on climate and the irreversibility of those effects makes global warming a uniquely dangerous and intractable problem for humanity." Unfortunately, the lag between cause and effect also makes it uniquely difficult to mobilize political will to address the problem, even though, as McKibben (2010, p. 27) notes, "The planet on which our civilization evolved no longer exists … The Earth that we knew-the only earth that we ever knew is [already] gone."

The environmental damage caused by global warming will result in a wide range of social, economic, and political harms to human communities and social systems on which they depend (Dyer 2010; Parenti 2011). For example, the rise in sea levels, extreme heat, and chronic droughts will lead to drastic reductions in the food supply, increasing famine, and mass migrations. The large movement of people across borders seeking food and an escape from the environmental consequences of increased temperatures will continue to fuel violent conflicts, genocides and other crimes. These movements have already led to the militarization and securitization of borders in the global North as neoliberal policies intersect with climate change to produce declines in agricultural and pastoral economies in the global South, leading to increasing mobility of irregular migrants seeking a minimum of food and/or physical security (Dunn 1995; Parenti 2011). Massive social upheavals, class conflict, and pandemics caused by climate change will stress social institutions, create ideological turmoil and generate political crises. The number of failed and failing states will increase as their incapacity to adapt to climate change increases poverty and violence around the globe, but particularly in those parts of the global South Parenti (2011) terms "the tropic of chaos." Resource wars and other forms of international conflict will increase and perhaps even provoke the use of nuclear weapons. Increased warfare would also sabotage the very planetary cooperation needed to reduce further global warming (Dyer 2010). From a moral–legal standpoint, one of the most disturbing elements of this process is that it harms most those living in geographic areas and countries that have contributed the least to the problem (Lynch and Stretesky 2010). The Northern, industrial nations have, in the words of Foster (2009, p. 243), accumulated a huge "ecological debt" toward Third World countries due to resource plundering and the infliction of environmental harms. But it is a debt that the global North, so far, has demonstrated little interest in paying.

State-Corporate Crime

Given these catastrophic scenarios, green criminology argues that criminologists treat the grave harms, both present and future, resulting from global warming and climate change as serious crimes warranting criminological analysis. As White (2011, p. 36) points out: "Climate change is arguably the most important issue, problem and trend in the world today and a key area of interest to eco-global criminology." As criminologists take up this issue, Lynch et al. (2010, p. 215) argue that the *state-corporate crime* approach "provides a useful tool for examining" the crimes related to global warming and climate change. Likewise, in his analysis of transnational environmental crime, White (2011, p. 13) also notes the importance of investigating the nation state as "… a major facilitator of harm in its own right, either on its own or in conjunction with specific sectional interests (such as particular transnational corporations)."

The concept of state-corporate crime, developed by Kramer and Michalowski (1990, 2006), refers to serious social harms that result from the interaction of political and economic organizations. The idea emerged out of the recognition that some organizational crimes are the collective product of the interaction between a business corporation and a state agency engaged in a joint endeavor. The concept of state-corporate crime seeks to breach the conceptual wall between economic crimes and political crimes in order to create a new lens through which we can examine the ways illegal acts and social injuries often emerge from intersections of economic and political power. As Michalowski and Kramer (2007, p. 201) note: "Contemporary social scientists have largely forgotten what our nineteenth century counterparts knew so well. There is neither economics nor politics; there is only political-economy."

State-corporate crime has been formally defined as "illegal or socially injurious actions that result from a mutually reinforcing interaction between (1) policies and/or practices in pursuit of the goals of one or more institutions of political governance and (2) policies and/or practices in pursuit of the goals of one or more institutions of economic production and distribution" (Michalowski and Kramer 2006, p. 15). As this definition makes clear, Michalowski and Kramer also propose to extend the scope of criminology beyond legal definitions, incorporating harmful social actions that violate neither criminal nor regulatory laws at the state level. While the concept of state-corporate crime could be applied to illegal or other socially injurious actions in societies ranging from private production systems to centrally planned political economies, most of the research to date has focused on state-corporate crimes within the private production system of US capitalism (Michalowski and Kramer 2006). State-corporate crimes within a global capitalist economy involve the active participation of two or more organizations, at least one of which is in the civil sector and one of which is in the state sector. The time has come to extend this framework to the study of global crimes of neoliberal capitalism, and in particular to the critical role of corporations and political states in both promoting the release of greenhouse gases and refusing to seriously address the resulting consequences of global warming and planetary climate change.

As a sensitizing concept the term state-corporate crime has three useful characteristics. First, it directs attention toward the way in which upper-world crime emerges at organizational intersections, in this case the intersection of institutions of accumulation and institutions of governance. In doing so, it foregrounds the ways in which many deviant organizational outcomes are not discreet acts of institutional wrongdoing, but rather the product of the relationships between different social institutions pursuing different goals and responding to different sets of pressures.

Second, it approaches the state as a nexus of *relationships* rather than a set of governmental institutional actors (Wonders and Solop 1993; Sassen 1993). This relational model directs us to examine the ways in which horizontal and vertical relationships between economic and political institutions contain powerful potentials for the production of illegal and other socially injurious actions. For example, US legislation designed to address global climate change, or more often the failure to pass legislation to address climate change, cannot be understood simply as a set of decisions taken by legislators pursuing varying political or organizational agendas. Rather, these outcomes are the product of long chains of relationships and conflicts among carbon-intensifying corporations (e.g., petroleum, coal, auto and auto related, highway construction), carbon-reducing industries (e.g., alternative energy, urban mass transit), political organizations of workers and communities dependent on these various industries (e.g., unions, Chambers of Commerce), environmental organizations, lobbying firms, banks, and other nodes of finance capital with a stake in environmental policy, along with a complex stew of regulatory bodies, advisory commissions, think tanks, and international governance and nongovernmental organizations. It is the flow of "information," data, money, and interpersonal linkages along these channels of power that constitutes the real operations of the US capitalist state. This relational approach provides a more nuanced understanding of the processes leading to deviant state actions than approaches that treat governments as closed systems, or locate the wrongdoing within individual decision makers operating within individual institutions.

Third, approaching the state-corporate context as a relational process directs analytic attention to the vertical relationships between different levels of organizational action in government and business. It asks us to be alert to three things. First, is the way particular individuals can, by their institutional movements and locations, shape flows of information, data and money through what Mills (1956) called the "circulation of elites." Second is the way in which standard operating procedures and cultures within institutions can facilitate or inhibit deviant organizational behavior (Vaughn 1996). And third, the way larger-scale political economic arrangements define the particular relationship between capital and the state (e.g., regulatory welfare state, neoliberal workfare state, state capitalism, etc.) and shape the opportunities and rewards for both socially harmful and socially responsible behaviors by individuals and organizations (Jessop 1991).

By examining these three levels we can recognize that political–economic arrangements are more than technical mechanisms for determining the relationship between state and capital. They also reflect and reproduce particular ideologies not just of the relationship between capital and state, but the relationship between capital and individuals and individuals and the state.

State-Corporate Crimes Related to Global Warming

We suggest that the harmful consequences of global warming and climate change are shaped in fundamental ways by four forms of state-corporate crime. Two of these forms concern failures of *mitigation*, that is, the need to drastically reduce the production of carbon dioxide and other greenhouse gases. The other two are failures of *adaptation*, the process of adjusting to or preparing to live with the effects of climate change that are either already underway or inevitable given the damage already done. In both cases, mitigation and adaptation, state corporate crimes can occur either through acts of commission or acts of omission.

Before proceeding, we wish to be explicit that we recognize that global warming and climate change are, in the broadest sense, the cumulative outcome of 200 years of industrialization, and particularly the rapid acceleration in fossil fuel consumption over the last century. Increasing consumption of industrial commodities, the underlying cause of climate change, has long been fueled by corporations seeking profit and states seeking popular legitimacy. The fetishism of commodities has been further accelerated by the intentional production of material desires among consumers (Ewen 2001). Thus, we are also aware that the expansion of material consumption, a key driver of development policies and industrial practices that have poisoned and are poisoning the planet, are broadly supported in nations of the global South as well as the global North. Although these macrorealities may themselves be crimes as Zerzan (2002) suggests, they are beyond the scope of our consideration here. However, even within these grand historical processes that arguably transcend approbation as crime, it is possible to discern concrete state–corporate relationships (1) that caused knowable and predictable harm, and (2) that could have avoided by state and capital managers who chose not to do so. It is this more specific arena of state-corporate crime to which we now turn.

Mitigation Failures: State Inaction and Climate Change Denial

The failure by individual states and the international community to undertake any serious efforts to mitigate global warming constitutes what Michalowski and Kramer (2006) term a *state facilitated* corporate crime, that is, an action or set of actions designed to enable corporate and state actors to pursue some pattern of harmful behavior.

Anthropogenic global warming stems from the production of heat-trapping greenhouse gases. In view of the extensive scientific evidence of the environmental and social harm resulting from emission-caused global warming, it would be reasonable to expect that the international political community and its member states would move immediately and aggressively to mitigate the production of greenhouse gases. Yet, to the contrary, many of the key corporate and state actors responsible for the greatest production of greenhouse gases have chosen to not only continue their current production practices, but in many cases have supported policies that will expand greenhouse gas production (see Chap. 4).

Some critical commentators have argued that the continued high levels of carbon dioxide emissions by corporations and the US military are crimes of corporate and state violence. As James Hansen puts it: "The trains carrying coal to power plants are death trains. Coal-fired power plants are factories of death" (quoted in Foster 2009, p. 21). In a similar vein, Sanders (2009, p. 22) says, "The military—that voracious vampire—produces enough greenhouse gases, by itself, to place the entire globe, with all its inhabitants large and small, in the most imminent danger of extinction."

That the emission of greenhouse-gases beyond levels scientifically determined to be low enough to avoid or slow global warming should be illegal is noncontroversial for most green criminologists. We take this one step further, arguing that the failure to reduce or mitigate the production of greenhouse gases and decarbonize the economy should be understood as a form of state-corporate crime. The failure of state officials to take effective and immediate actions to compel both the private sector and governmental institutions to reduce their emission of greenhouse gases is, arguably, a state crime of omission. For example, the George W. Bush administration not only refused to pass domestic policies that would limit the production of greenhouse gases in the United States, but also worked proactively to sabotage the Kyoto Accord by effectively withdrawing from the Kyoto process and promoting its own strategy to address climate change, one which, would of course, have no impact on US business-as-usual. His acts were so egregiously supportive of industries and practices contributing to climate change that Kennedy (2004) accused the administration of "crimes against nature." As Lynch et al. (2010) document, 8 critical years were lost in the battle against global warming due to the stonewalling and negligence of an administration that had extraordinary ties to the main culprits in the coal and gas industry.

The United States is not alone in its foot-dragging on the matter of climate change. Other states, and the international political system in general, have also failed to take aggressive action. The Copenhagen Conference in December of 2009 may have been the last, best opportunity for the governments of the world to act forcefully to prevent catastrophic climate change. Yet, the Obama administration and the entire international political community failed to take any strong actions that might avert impending ecocide. As John Sauven, executive director of Greenpeace U.K. bluntly stated after the failure of the conference: "The city of Copenhagen is a crime scene tonight, with the guilty men and women fleeing to the airport. There are no targets for carbon cuts and no agreement on a legally binding treaty" (BBC 2009, p. 3). Similarly, White (2011, p. 148) contends that the failure at the Copenhagen conference was indeed a state-corporate crime, noting that "The abject failure of the Copenhagen talks to actually do something about carbon emissions and to address climate change issues in a substantive fashion is a striking example of the fusion of state and corporate interests to the detriment of the majority."

While the failure to take aggressive action to limit global warming is a state-corporate crime of omission in our schema, the orchestrated denial of climate change, despite extensive evidence to the contrary, is a –state-corporate crime of commission. It is not a failure to act, but a deliberate attempt to thwart efforts to respond in an effective and just way to the emerging problems resulting from global warming.

Global warming denial efforts are largely carried out by conservative think tanks funded for the most part by money from the fossil fuel industry (Gelbspan 2004; Greenpeace 2011; Jacques et al. 2008; Oreskes and Conway 2010). For example,

Western Fuels, a large coal cooperative, and the giant Exxon Mobil oil company, have each contributed millions of dollars to conservative think tanks and environmental skeptics working to deny global warming (Adams 2009; Jacques 2009; McNall 2011). Oreskes and Conway (2010, p. 247) note that, "Exxon Mobil's support for doubt-mongering and disinformation is disturbing but hardly surprising. What is surprising is to discover how extensive, organized, and interconnected these efforts have been, and for how long."

The global warming denial countermovement consists largely of corporate propaganda built around lies and deceptions masquerading as science which is then disseminated through conservative think tanks, industry trade associations, right-wing opinion leaders, the corporate media and by some government officials (Friel 2010; Gelbspan 1998, 2004; Hamilton 2010; Hoggan 2009; Jacques 2009; McCright and Dunlap 2000, 2003; Oreskes and Conway 2010; Chap. 4). The intent of these denial efforts is to cast doubt on the scientific evidence of anthropogenic global warming, and thus impede governmental actions that would force the fossil fuel industry to make changes that would reduce emissions. Some environmental scientists contend that such corporate and/ or government-sponsored climate science disinformation and denial should be labeled as crime. For instance, climate scientist Donald Brown (2010, p. 2) states, "We may not have a word for this type of crime yet, but the international community should find a way of classifying extraordinarily irresponsible scientific claims that could lead to mass suffering as some type of crime against humanity."

While space limitations preclude an extended analysis, it is important to at least note that state-corporate crimes that result in catastrophic climate change are rooted in broader structural and cultural forces. The continued "criminal" emission of greenhouse gases in much of the world arises from the global dominance of a predatory corporate capitalist economic system and the popular desires it stimulates, protected by imperial economic and/or military actions against any nations that might seek to pursue policies and practices that contradict the interests of that system (Burbach et al. 1996). As Foster (2009, p. 46) points out, "Capital by its very logic imposes what is in effect a scorched earth strategy. The planetary ecological crisis is increasingly all-encompassing, a product of the destructive uncontrollability of a rapidly globalizing capitalist economy, which knows no law other than its own drive to expand."

The global political economy of predatory capitalism also gives rise to two broad cultural factors that, in turn, reinforce the economic relationships and forces that facilitate global warming. The two factors are "growth fetishism" (Hamilton 2010) and state supported cultures of consumption and production (Ewen and Ewen 1982; Lynch and Stretesky 2010). The pathological promotion and pursuit of endless economic growth on a planet with finite resources, such as fossil fuels, is unsustainable in the long run. It also produces "tunnel vision" which restricts people from considering any solution to the global warming problem other than a technological one which would facilitate continued high levels of consumption, only at a "cleaner" level (Hamilton 2010). Insofar as "The more an individual or culture consumes, the more that person or culture contributes to climate change" (Lynch and Stretesky

2010, p. 64), the only path away from continued global warming is reductions in levels of consumption. Unfortunately, any US political leader (or probably the leader of any other nation of the North) who suggests individuals consider lives built around consuming less does so at grave political risk. The only US President to make such an effort was Jimmy Carter who, in the midst of the fuel crisis of the 1970s, suggested Americans lower their thermostats and put on sweaters. He was pilloried for the very idea that Americans should use less (Carter 1977). Yet, had the United States pursued his overall energy policy of reduced consumption, its carbon footprint would be less today than it is.

Jacques (2009) also analyzes both structural and cultural factors behind environmental skepticism in general and climate change denial specifically. He argues that environmental skepticism is a social countermovement organized by corporate funded conservative think tanks. These think tanks provide cover for private industry and the conservative ideology itself. According to Jacques (2009, p. 45), environmental skepticism "is working to counter the advances of international diplomacy and negotiations about trans-boundary environmental changes," such as climate change. He argues that this broad based countermovement is rooted in and seeks to protect the world capitalist system, or what Hippwell (2004, p. 370) calls *Industria*, an "industrial, homogenizing force" that manifests itself "as an anthropocentric, rationalizing, colonizing and ecologically destructive network of capture and control."

The environmental skepticism behind climate change denial is more complex than simply protecting profits or distributional interests. As Jacques (2009, p. 89) argues, the skeptical world-view is "held together by a deep anthropocentrism that seeks to annihilate non-human ecology or at least has little ethical use for non-humans, a severely narrow sense of civic obligation and duty, and an ontology of possessive individualism that sees consumption and property as defining features for being human." This anthropocentric view is directly counter to what pioneering environmentalist Aldo Leopold (1949/1989, p. 204) termed a "land ethic," which he describes as a way of thinking and being that "simply enlarges the boundaries of [our] community to include soils, waters, plants, and animals, or collectively: the land." In contrast to the environmental-domination perspective that Jacques critiques, a land ethic asks that we behave in ways that "affirm" and protect the right of the land "to continued existence, and, at least in spots … continued existence in a natural state."

These two perspectives reveal that at the root of the climate-change debate lays a much deeper conflict over fundamental visions of the relationship of humans to their environment, and by extension, ultimately to one another. The environmental-domination view is essentially modernist and closely associated with the rise of accumulation as the central engine of political economic organization, whether that accumulation be through private capitalism or some form of state capitalism. The land-ethic standpoint is simultaneously pre-modern and future-modern. That is, it both reflects the practices of the pre-modern era when humans fully recognized their dependence on the land and thus treated it in ways to ensure their survival, and (possibly) the ideology and practices of a future period when we once again share the recognition of this intimate connection with the land, and thus our need to preserve it and all that it contains for both human survival and for the equally valid good of protecting "the land" for its own sake (Zerzan 2002).

Unfortunately, the human-centric and modernist vision of environmental domination not only serves the economic and political interests associated with neoliberal capitalism's agenda to exploit planetary resources to the maximum for profit, but they also intersect with powerful cultural forces that believe human consumption of commodities should take precedence over environmental protection. Combined, these forces help support a climate-change denial movement that facilitates the continued production of greenhouse gases and, by sowing seeds of doubt about global warming, helps block regulative efforts on behalf of mitigation.

The Politics of the Armed Lifeboat and the Exclusion of Progressive Political Adaptation

Many environmental activists resist even talking about adaptation (actions designed to reduce vulnerability to the negative effects of climate change), insisting that the mitigation of global warming must be placed first and foremost on the action agenda. But as McKibben (2010) has pointed out, due to climatic changes that have occurred, we live on a planet today that is already significantly different from what it was at the end of World War II. Those who accept that climate change is already underway, argue that it is imperative we explore the least destructive ways to adapt to these changes while simultaneously seeking to mitigate the causes of global warming.

Adaptation can take two forms. One form is positive, progressive, cooperative and socially just. The other is militarized and repressive, what Parenti (2011) calls the "politics of the armed lifeboat." We contend that the exclusion of the first type of political adaptation to climate change from economic and political discourse and the adoption of the second, are state-corporate crimes insofar as they will bring predictable and avoidable harm to large portions of the human population in order to benefit smaller segments in the richest and most powerful nations of the world.

Hertsgaard (2011) describes a number of adaptations to climate change already underway. Some local governments, such as in Seattle, Chicago and New York are responding with strategies that simultaneously serve mitigation and adaptation by seeking routes to sustainable economic development, increasing energy efficiency, planting more trees and shifting to wind powered electricity. He points out that as an adaptation to the threat of rising ocean levels the Netherlands is in the process of raising the height of its sea walls. A tree-based approach to farming, called "farmer-managed natural regeneration" has transformed the western Sahel (the climatic band between the Saharan and savanna areas of Africa) in recent years. Other ecological agricultural practices to increase food production are also being investigated (Hertsgaard 2011).

The problem with such measures is that there are too few of them, they are localized and widely scattered, lack political support, and are often underfunded.

Moreover, many of these examples do not address the significantly greater impact global climate change will have on the human populations of less developed countries. According to the IPCC (2001, 2007), climate change will be far more devastating for populations in less developed countries insofar as they depend more heavily on the environment for subsistence, already face problems of food insecurity, desertification, limited access to potable water, often have low levels of arable land relative to population, and lack levels of technological development that might be adapted to ameliorate the impacts of climate change.

The consequences of global climate change in less developed countries will reach well beyond the boundaries of devastated areas. As Reuveny (2007, p. 656) notes:

> People can adapt to environmental problems in three ways: stay in place and do nothing, accepting the costs; stay in place and mitigate changes; or leave affected areas. The choice between these options depends on the extent of the problems and mitigation capabilities. Developed countries (DCs) are likely to mitigate problems through technological innovation and institutional redesign. Less developed countries (LDCs) are less likely to mitigate such problems since they lack wealth and expertise.

Insofar as human populations typically do not accept their demise passively, we can anticipate substantial climate induced migration from less developed countries as the effects of global climate change deepen. Over two decades ago the IPCC (1990, p. 2) warned that the "greatest single impact of climate change could be on human migration with millions of people displaced by shoreline erosion, coastal flooding and agricultural disruption." More recently a report by the Asian Development Bank (2009) concluded that in the Asia/Pacific region alone anywhere from 700 million to one billion people "will come under substantial pressure to migrate (temporarily or permanently, and internally or across borders)" due to climatological disruptions to shorelines and food systems.

These migrations will not be benign. In addition to the deep disruption to the lives of those who are forced or feel compelled to migrate due to climate induced environmental changes, these migrations hold a significant threat of violent conflict. According to estimates by Reuveny (2007), between 1960 and 1990 there were 36 violent conflicts resulting from or exacerbated by climate induced migration due to increased competition for resources, intensified ethnic tensions, inter- and intra-governmental distrust and deepening socio-political fault lines. These data cover only the earliest possible impacts of global climate change. They suggest that in the absence of genuinely progressive, cooperative adaptations to global climate change, as the impacts of climate deepen, we can expect a significant increase in migration induced conflicts.

Despite these looming threats, adaptation to global climate change was not even on the agenda of the international political community until the developing nations of the global South demanded adaptation assistance and funding from the rich nations at the Copenhagen conference. This, however, became a significant stumbling block to an agreement because governments of the global North refused to acknowledge that they owed an ecological dept to the South, let alone act to reduce this debt. The states (and corporations) that derive the most benefits from the global capitalist economy have refused to participate in any adaptation efforts that require "economic redistribution and development" or "a new diplomacy of peace building" (Parenti 2011, p. 11). Given the likelihood of violence resulting from climate-induced migration, the failure to plan for and adopt progressive, cooperative and just adaptation policies warrants analysis of global climate change as a state-corporate crime of omission.

While the failure to adopt peaceful measures of adaptation is a crime of omission, the state-corporate nexus is also guilty of the direct commission of a violent crime by "responding to climate change by arming, excluding, forgetting, repressing, policing, and killing" (Parenti 2011, p. 11). Parenti (2011, p. 7) notes that, "The current and impending dislocations of climate change intersect with the already-existing crises of poverty and violence," crises that are the products of Cold War politics and neoliberal economic policies. In his words, this collision of global poverty and violence with climate change constitutes "the catastrophic convergence." States in the global North are responding to this catastrophic convergence primarily with militarism, violence and repression. As Parenti (2011, p. 11) observes:

> One can imagine a green authoritarianism emerging in the rich countries, while the climate crisis pushes the Third World into chaos. Already, as climate change fuels violence in the form of crime, repression, civil unrest, war, and even state collapse in the Global South, the North is responding with a new authoritarianism. The Pentagon and its European allies are actively planning a militarized adaptation, which emphasizes the long-term, open-ended containment of failed or failing states-counterinsurgency forever.

Parenti (2011, p. 11) adds, "This sort of 'climate fascism,' a politics based on exclusion, segregation, and repression, is horrific and bound to fail." We contend that a militarized response to climate conflicts, such as the 2006 US proxy war and other forms of military involvement in drought and famine stricken Somalia (Scahill 2011), is state-corporate crime designed to keep powerful economies and their governments in power at the expense of the rest of the world. Whether we have reached the point of catastrophe as Parenti suggests, or whether there is still time to avert its worst consequences, the path toward some alternative and less grim future lies through important transformations at the intersection of corporations and states, and it is here that a public criminology can play an affirmative role by confronting the state-corporate crimes of climate change.

Conclusion: Public Criminology, State-Corporate Crime, and Climate Change

This chapter has argued that global warming is a state-corporate crime that warrants further criminological inquiry. We end with a plea for the development of a public criminology of the crimes related to global warming and climate change. Following Burawoy's (2007) conceptualization of public sociology, Kramer et al. (2010) have argued that a public criminology of state crime would seek out extra-academic audiences and enter into conversations with various publics concerning these crimes.

Public criminologists willing to speak about global warming and climate change in what Jensen (2009) calls the "prophetic voice," then have a responsibility to act in the public arena. Kramer (2012) has recently offered three ways that criminologists can engage in progressive political action to confront the state-corporate crimes analyzed in this chapter. The first approach is to play a role in breaking through the denial and normalization that usually covers crimes related to global warming [presenting research to document climate harms and dispute literal denials of these crimes or theory to counter narratives of interpretive denial]. The second involves engaging in transnational activism with social movement groups to contest the power of the corporate state in an effort to achieve specific progressive policy changes concerning the de-carbonization of the global economy and progressive adaptations to climate change [acting as consultants to specific environmental NGOs such as Greenpeace or Climate Action Network who are working to challenge corporate agendas]. Finally, Kramer argues that criminologists can contribute their insights and understanding to help enhance the ability of international legal institutions to establish controls over global warming [advocating for international agreements that cut greenhouse gas emissions and promote the development of alternative energy sources].

Pursuing these tasks is not easy. They require time and energy which many academic criminologists already find in short supply. These tasks also require transcending the "normal science" mandate that researchers and scholars refrain from activism that would undermine their "value neutrality." However, as Robert Heilbroner (1974) once observed, when it comes to social inquiry into "the human prospect," value neutrality is always an impossibility. Rather than simply be observers cataloguing state-corporate crimes, criminologists concerned with climate change need to engage as public intellectuals, that is, as overt activists for new visions of how humans can live on this planet and how economic and political institutions can be remade in pursuit of those new visions.

References

Adams, D. (2009, July 1). ExxonMobil continuing to fund climate skeptic groups, records show. *The Guardian*.

Asian Development Bank. (2009). *Policy options to support climate-induced migration*. Research and Development Technical Assistance (RDTA) Project Number: 43181-01. http://www.adb.org/Documents/TARs/REG/43181-REG-TAR.pdf.

Barak, G. (Ed.). (1991). *Crimes by the capitalist state: An introduction to state criminality*. Albany: State University of New York Press.

BBC. (2009, December 19). Climate talks battle to save deal. *BBC News*. http://news.bbc.co.uk/2/hi/science/nature/8422031.stm.

Beirne, P., & South, N. (Eds.). (2006). *Green criminology*. Aldershot: Ashgate.

Beirne, P., & South, N. (Eds.). (2007). *Issues in green criminology: Confronting harms against environments, other animals and humanity*. Cullompton: Willan Publishing.

Broder, J. (2010, November 8). A novel tactic in climate fight gains some traction. *The New York Times*. http://www.nytimes.com.

Brown, D. (2010, November 3). Is climate science disinformation a crime against humanity? *Commondreams*. http://www.commondreams.org.

Burawoy, M. (2007). For public sociology. In D. Clawson, R. Zussman, J. Misra, N. Gerstel, R. Stokes, D. Anderton, & M. Burawoy (Eds.), *Public sociology: Fifteen eminent sociologists debate politics and the profession in the twenty-first century* (pp. 23–64). Berkeley: University of California Press.

Burbach, R., Nunez, O., & Kagarlitsky, B. (1996). *Globalization and its discontents: The rise of postmodern socialisms*. London: Pluto Press.

BusinessGreen. (2011a). US states accuse polluting power stations of 'public nuisance'. *BusinessGreen*. http://www.businessgreen.com/bg/news/2034401/accuse-polluting-power-stations-public-nuisance.

BusinessGreen. (2011b). Green groups to sue US government over atmospheric pollution. *BusinessGreen*. http://www.businessgreen.com/article/bg/news/2068121/green-sue-government-atmospheric-pollution.

Carter, J. (1977). Proposed energy policy. http://www.pbs.org/wgbh/americanexperience/ features/primary-resources/carter-energy/.

Dunn, T. (1995). *Militarization of the U.S.-Mexico border: Low intensity warfare comes home*. Austin, TX: University of Texas Press.

Dyer, G. (2010). *Climate wars: The fight for survival as the world overheats*. Oxford: Oneworld Publications.

Ewen, S. (2001). *Captains of consciousness: Advertising and the social roots of consumer culture*. New York: Basic Books.

Ewen, S., & Ewen, E. (1982). *Channels of desire: Mass images and the shaping of American consciousness*. New York: McGraw-Hill.

Foster, J. B. (2009). *The ecological revolution: Making peace with the planet*. New York: Monthly Review Press.

Friel, H. (2010). *The Lomborg deception: Setting the record straight about global warming*. New Haven: Yale University Press.

Gelbspan, R. (1998). *The heat is on: The climate crisis, the cover-up, the prescription* (Updated Edition). Reading, MA: Perseus Books.

Gelbspan, R. (2004). *Boiling point: How politicians, big oil and coal, journalists, and activists are fueling the climate crisis-and what we can do to avert disaster*. New York: Basic Books.

Greenpeace. (2011). *Who's holding us back? How carbon intensive industry is preventing effective climate legislation*. Amsterdam: Greenpeace International.

Hamilton, C. (2010). *Requiem for a species: Why we resist the truth about climate change*. London: Earthscan.

Hansen, J. (2009). *Storms of my grandchildren: The truth about the coming climate catastrophe and our last chance to save humanity*. New York: Bloomsbury.

Heilbroner, R. (1974). *An inquiry into the human prospect*. New York: W.W. Norton.

Hertsgaard, M. (2011). *Hot: Living through the next fifty years on earth*. Boston: Houghton Mifflin Harcourt.

Higgins, P. (2010). *Eradicating ecocide: Exposing the corporate and political practices destroying the planet and proposing the laws needed to eradicate ecocide*. London: Shepheard-Walwyn Publishers.

Hillyard, P., Pantazis, C., Tombs, S., & Gordon, D. (Eds.). (2004). *Beyond criminology: Taking harm seriously*. London: Pluto Press.

Hippwell, W. T. (2004). A Deleuzian critique of resource-use management politics in industria. *The Canadian Geographer, 48*, 356–377.

Hoggan, J. (2009). *Climate cover-up*. Vancouver: Greystone Books.

IPCC. (1990). *Climate change: The IPCC impacts assessment report*. Canberra: Australian Government Publishing Service. Prepared for Intergovernmental Panel on Climate Change by Working Group II. Canberra.

IPCC. (2001). *Special report on the regional impacts of climate change: An assessment of vulnerability*. Geneva: Intergovernmental Panel on Climate Change.

IPCC. (2007). *Climate change 2007: Climate change impacts, adaptation, and vulnerability*. Geneva: Intergovernmental Panel on Climate Change.

Jacques, P. (2009). *Environmental skepticism: Ecology, power and public life*. Farnham: Ashgate.

Jacques, P., Dunlap, R., & Freeman, M. (2008). The organization of denial: Conservative think tanks and environmental skepticism. *Environmental Politics, 17* (June), 349–385.

Jensen, R. (2009). *All my bones shake: Seeking a progressive path to the prophetic voice*. Berkeley: Soft Skull Press.

Jessop, B. (1991). *State theory: Putting capitalist states in their place*. Philadelphia: University of Pennsylvania Press.

Kauzlarich, D., Mullins, C., & Matthews, R. (2003). A complicity continuum of state crime. *Contemporary Justice Review, 6*, 241–254.

Kennedy, R. F. (2004). *Crimes against nature*. New York: HarperCollins.

Kramer, R. (2012). Public criminology and the responsibility to speak in the prophetic voice concerning global warming. In E. Stanley & J. McCulloch (Eds.), *State crime and resistance*. London: Routledge. forthcoming.

Kramer, R., & Michalowski, R. (1990, November). *Toward an integrated theory of state-corporate crime*. Paper presented at the American Society of Criminology Meeting, Baltimore, MD.

Kramer, R., & Michalowski, R. (2006). The original formulation. In R. Michalowski & R. Kramer (Eds.), *State-corporate crime: Wrongdoing at the intersection of business and government* (pp. 18–26). New Brunswick, NJ: Rutgers University Press.

Kramer, R., Michalowski, R., & Chambliss, W. (2010). Epilogue: Toward a public criminology of state crime. In W. Chambliss, R. Michalowski, & R. Kramer (Eds.), *State crime in the global age* (pp. 247–261). Cullompton: Willan Publishing.

Leopold, A. (1949). *A sand county almanac and sketches here and there* (Reissued 1989). New York: Oxford University Press.

Lynch, M., Burns, R., & Stretesky, P. (2010). Global warming and state-corporate crime: The politicalization of global warming under the Bush administration. *Crime, Law and Social Change, 54*, 213–239.

Lynch, M., & Stretesky, P. (2003). The meaning of green: Contrasting criminological perspectives. *Theoretical Criminology, 7*, 217–238.

Lynch, M., & Stretesky, P. (2010). Global warming, global crime: A green criminological perspective. In R. White (Ed.), *Global environmental harm: Criminological perspectives* (pp. 62–84). Cullompton: Willan Publishing.

McCright, A., & Dunlap, R. (2000). Challenging global warming as a social problem: An analysis of the conservative movement's counter-claims. *Social Problems, 47*, 499–522.

McCright, A., & Dunlap, R. (2003). Defeating Kyoto: The Conservative Movement's impact on U.S. climate change policy. *Social Problems, 50*, 348–373.

McKibben, B. (2010). *Eaarth: Making life on a tough new planet*. New York: Times Books.

McNall, S. (2011). *Rapid climate change: Causes, consequences, and solutions*. New York: Routledge.

Michalowski, R. (1985). *Order, law and crime*. New York: Random House.

Michalowski, R. (2010). In search of 'state and crime' in state crime studies. In W. Chambliss, R. Michalowski, & R. Kramer (Eds.), *State crime in the global age* (pp. 13–30). Cullompton: Willan Publishing.

Michalowski, R., & Kramer, R. (2006). *State-corporate crime: Wrongdoing at the intersection of business and government*. New Brunswick, NJ: Rutgers University Press.

Michalowski, R., & Kramer, R. (2007). State-corporate crime and criminological inquiry. In H. Pontell & G. Geis (Eds.), *International handbook of white-collar and corporate crime* (pp. 200–219). New York: Springer.

Mills, C. W. (1956). *The power elite*. Oxford: New York.

Oreskes, N., & Conway, E. (2010). *Merchants of doubt: How a handful of scientists obscured the truth on issues from tobacco to global warming*. New York: Bloomsbury Press.

Parenti, C. (2011). *Tropic of chaos: Climate change and the new geography of violence*. New York: Nation Books.

Pemberton, S. (2007). Social harm future(s): Exploring the potential of the social harm approach. *Crime, Law and Social Change, 48* (1–2), 27–41.
Reuveny, R. (2007). Climate change-induced migration and violent conflict. *Political Geography, 26*, 656–673.
Richardson, K., Steffan, W., & Liverman, D. (2011). *Climate change: Global risks, challenges and decisions*. New York: Cambridge University Press.
Sanders, B. (2009). *The green zone: The environmental costs of militarism*. Oakland: AK Press.
Sassen, S. (1993). *Losing control? Sovereignty in the age of globalization*. New York: Columbia University Press.
Scahill, J. (2011). Blowback in Somalia: How US proxy wars helped create a militant Islamist threat. *The Nation, 293*, 11–22.
Tift, L., & Sullivan, D. (2001). A needs-based, social harms approach to defining crime. In S. Henry & M. Lanier (Eds.), *What is crime? Controversies over the nature of crime and what to do about it* (pp. 179–203). Lanham, MD: Rowman & Littlefield.
Vaughn, D. (1996). *The challenger launch decision: Risky technology, culture and deviance at NASA*. Chicago: University of Chicago Press.
White, R. (2008). *Crimes against nature: Environmental criminology and ecological justice*. Cullompton: Willan Publishing.
White, R. (Ed.). (2010). *Global environmental harm: Criminological perspectives*. Cullompton: Willan Publishing.
White, R. (2011). *Transnational environmental crime: Toward an eco-global criminology*. London: Routledge.
Wonders, N., & Solop, F. (1993). Understanding the emergence of law and public policy: Toward a relational model of the state. In W. Chambliss & M. Zatz (Eds.), *Making law: The state, the law and structural contradictions* (pp. 204–228). Bloomington, IN: University of Indiana Press.
Zerzan, J. (2002). *Running on emptiness: The pathology of civilization*. Los Angeles: Feral House.

Chapter 6
Climate Change in the Courts: A US and Global Perspective

Andrew Franz

Climate Change: Why the Courts?

Global temperatures continue to rise—year after year new heat records are set; the permafrost is melting and rotting—leading to damaging summer methane emissions; glaciers are receding everywhere; species are endangered—perhaps a quarter of species now on earth will be extinct by mid-century, and half by 2100; incidents of drought, poor water quality, crop losses, landslides, pest inundation, severe storms, raging wildfires, and tropical diseases are increasing and spreading; massive human dislocations—particularly in low-lying coastal areas inundated by rising seas—are already occurring and expected to get worse; deforestation—which also drives climate change—continues at ever rapid pace in the world's great jungles; in the summer of 2004–2005, the oceans turned from carbon sink to carbon producers (Sinden 2007; Wood 2007; Irwin 2010). We are witnessing the "end of nature" where we have reached a "tipping point" such that the "feedback loops" have been triggered and devastating consequences are unavoidable. The scientific community is certain to a very high degree that the cause of these harms is anthropogenic (McKibben 1989; Sinden 2007; IPCC 2007). The anthropogenic impact on Earth has grown so massive that we have come to an epochal moment (Irwin 2010).

While all of humanity and nature are at dire risk, clearly some are more culpable for this being so, and some are more vulnerable because of it. Wealthy countries such as the USA, which accounts for nearly 30% of greenhouse gas emissions (GHGs), are responsible for the vast majority of present and historically accumulated climate change activity (Sinden 2007; Wood 2007). The vast majority of GHG emissions are caused by energy and transportation related activities, many of them

A. Franz (✉)
Department of Criminology, University of South Florida, Tampa, FL 33620-8100, USA

R. White (ed.), Climate Change from a Criminological Perspective, DOI 10.1007/978-1-4614-3640-9_6,

transnational in nature, all transnational in effect. Profound moral issues implicating widespread denial of human [and animal] rights demand a profound response from the law (Sinden 2007). Government's climate obligations toward future generations are great, yet we lack legal guidance in this time of crisis. For rights implicated by climate change to be considered efficacious they must be subject to enforcement mechanisms in the courts that are fair and open. Climate change involves market failure, political failure, scientific failure, and social failure, and all of these in turn lead to rights violations. The courts are where such power imbalances and institutional failures traditionally are vindicated, and tellingly, the vast majority of climate change cases have been brought, not by corporations, but by peoples and non-governmental organizations (NGOs) who have been injured by industry externalities (Osofsky 2006).

In order to understand what kind of job the courts are doing to rectify the harms of climate change this chapter is going to first explore the litigation situation in the USA. This is because the USA has the greatest per capita carbon emissions in the world; its pollution and policies impact the world more than any other nation; and it has stood in the way of a number of international conventions and protocols. Next, the chapter looks at foreign and international courts. These courts, because they often intersect, are difficult to categorize and treated in a more summary fashion, but the sum total of activity occurring in them is nonetheless very important. Throughout the chapter and beyond the legalese, the issues of rationalism, science, economics, politics, the role of administrative agencies, the concept of "environmental justice" and the field of criminology will form part of the discussion. The question that the reader should ask throughout is this: "given the facts, what would the most rational or the best legal system look like in regard to the problems posed by climate change?"

The US Courts' Response to Climate Change

In the US federal courts, there have been a handful of major cases in the regular courts on the issue of climate change. With few exceptions the federal courts have not been friendly to those seeking redress in them for the harms of carbon dioxide related climate change. On the other hand, chemical causes of climate change such as chlorofluorocarbons (CFCs), hydrochlorofluorocarbons, methyl chloroform, and methyl bromide, despite preliminary resistance to regulation by industry and reports of a vigorous black market for materials, now seem more settled in law, e.g., *Northwest Environmental Defense Center v. Owens Corning Corp.*, 434F. Supp. 2d 957 (D. Or. 2006) (finding standing for Oregonians to sue an Oregon facility over permits to emit ozone-depleting substances in Oregon). Methane, which also strongly contributes to climate change, is much more likely to be litigated in the near future, especially with the recent discovery that increasingly popular "shale gas fracking" releases some 20 times more methane than originally thought which, in the shorter run (5–10 years), is over 100 times worse for the climate than carbon dioxide. The US federal courts, consisting of District Courts (trial level), Circuit Courts (intermediate appeal level), and a single Supreme Court (the ultimate and final reviewing judicial body), have widely indicated that they prefer to remain out of the climate change debate. These courts have maintained their distance from the climate change debate by determining one or more of either (a) claimants lack "standing" (e.g., an injured party's right to make a legal claim or seek judicial enforcement of a statutory duty or a constitutional right) or (b) claimants cannot show that their injury was "proximately caused" (e.g., a cause legally sufficient to produce liability for an event complained of) by defendant's conduct or (c) the matter is a "political question" better handled by an exercise of discretion by another branch of government (e.g., the executive or legislature) instead of the courts.

If history is any predictor, as the impacts of climate change grow worse and more obviously present a threat to life on earth, this legal posture will likely change in the future as did analogous situations in the past (slavery, tobacco, etc.). That these changes will occur in time to respond to the exigencies, however, is doubtful. Traditionally, the federal courts have responded, albeit carefully, to politically untenable or insular situations such as discrimination, or privacy. Given the federal courts' recalcitrance so far, as well as the political climate, it seems likely that the real change in US climate policy will occur piecemeal, by challenging permits and regulations at the state and federal level [for autos, power plants, and manufacturing (Mounteer 2009)],within administrations, through international pressures, public opinion shifts (Americans do like to have peace of mind over future uncertainty and demonstrate a willingness to pay for disaster avoidance—but see Chap. 4), and examples brought home from international tribunals.

At the state level, it is notable there have been over 50 climate change related suits, many very recently filed. Most of these have occurred in California, and they generally pertain to either (1) the National Environmental Policy Act (2000) (NEPA) laws (e.g., failure to do Environmental Impact Assessments) or (2) reasonableness of agency permitting or (3) state and local vehicle emission regulations (Gerrard 2007; Osofsky 2006). While the number of these suits continues to grow and do seem to form an effective sort of strategy (see, for instance: http://www.climate-casechart.com), still, in most states, climate change issues are best described as a "race to the bottom." County commissioners continue to approve trophy home developments, permits for GHG emitting industries are handed out as if climate change were a myth, forestry services continue to deliver massive timber sales, and large numbers of permits are still sought for the worst GHG offender—coal burning power plants (Wood 2007).

The primary cases in the US federal courts are two Supreme Court cases: *Massachusetts v. Environmental Protection Agency* (EPA), 549 US 497 (2007) (hereinafter: *Massachusetts*); and *American Electric Power Co., Inc. v. Connecticut*, Slip Opinion No. 10-174, 20 June 2011 (hereinafter, "*AEP*").

The *Massachusetts* case was brought by 12—mostly northeastern—states, three large eastern cities, a US territory, and a number of NGOs. Defendants included the EPA, car and truck manufacturing associations, two industry NGOs, and ten states—mostly from the Midwest and oil producing states. The facts of the case were that the EPA had refused to regulate automobile emissions for GHGs despite mounting pressure to do so, and that the plaintiffs sued to compel agency action. The intermediate appellate court, in this instance the first line of review in the regular courts, held that the EPA had discretion not to regulate GHGs, and dismissed the case for lack of standing due to what the judges called "scientific uncertainty" and the plaintiff's inability to show that they had been "peculiarly harmed" by vehicle emissions since, that court opined, global warming was a global, not a regional, issue.

The Supreme Court reversed the Circuit Court, finding that of the claimants, at least, the 12 states had standing since states are charged with overseeing the public health and welfare of their citizens in the face of the potentially enormous harm of global warming. The high court also found that despite the fact that the Clean Air Act (CAA) never really mentioned Climate Change when written in 1970 or amended in 1977 and 1990, GHGs fit well within the broad definition of "air pollutants" which, for purposes of "public welfare," must be regulated by EPA. This meant that EPA, though a careful reading of the facts indicated they likely suffered "agency capture," could not continue to claim it lacked authority to regulate GHG emissions for new vehicles or that it, in the alternative, would be unwise to do so. The decision was a mere 5–4, but for the first time the Court recognized the science that carbon-dioxide emissions from autos cause climate change, which thereby damages the environment in a number of ways (Mounteer 2009; Reeves 2009). For instance, the majority opinion spoke of increases in heat-related deaths; coastal flooding and erosion caused by melting icecaps and rising sea levels; more frequent, intense and extreme weather events, such as hurricanes, drought, and precipitation patterns that cause death, disrupt food production, destroy infrastructure, as well as plant and animal ecosystems. The decision, however, did not broach whether it applied only to regulating motor vehicles or would be expanded into other areas where GHG emissions occur. The conservative block on the Court dissented from the holding, with the Chief Justice expressing his "formalistic" and "inconsequentialist" fears the decision had opened the floodgates to environmental litigation, which he believed should have narrow standing grounds due to harms whose origins are hard to trace or be redressed other than politically.

A "sort of" clarification came four years later in *AEP* where a number of states, cities, and NGOs sued five major electric power companies for substantially and unreasonably causing the federal and state "public nuisance" (e.g., a condition dangerous to health, or that unlawfully obstructs the public in the free use of public property. An unreasonable interference with a right, common to the general public and behavior which unreasonably interferes with the health, safety, peace, comfort, or convenience of the general community) of global warming (see also, Chap. 5). Prior to this case, NGOs had found it difficult to sue power plants, called "stationary sources," under the *Massachusetts* holding—which had mandated a regulatory, not a civil tort suit, approach (Mounteer 2009). In *AEP* defendants were alleged to be the five largest emitters of carbon dioxide in the USA, thereby accounting for 2.5% of all anthropogenic emissions worldwide. Claiming public lands, infrastructure and health were at risk, the plaintiffs sought an injunctive decree that would cap and subsequently reduce defendants' power plant carbon-dioxide emissions. By this time, EPA had initiated but not completed administrative rule making, not just for new automobiles but also for new, modified, and existing power plants. The federal trial court dismissed the complaint on the grounds that it presented a "non-justiciable political question," but the intermediate appeals court reversed, holding plaintiffs were not barred by the political question doctrine, and that they had standing.

The Supreme Court took the case, and unanimously reversed the intermediate appellate court's holding. Though it admitted public nuisance law is flexible and adapts to changing science and facts, the Supreme Court seemed to withdraw from its position in the *Massachusetts* case that climate change was caused by carbon dioxide or even that it was harmful. The high court cautioned in a footnote, "The Court… endorses no particular view of the complicated issues related to carbon dioxide emissions and climate change." Importantly, though it overruled the Circuit Court's decision the Supreme Court did not, however, overrule the part of the prior holding that found the claimants were not barred by the political question doctrine. The grounds for overruling the appellate court and dismissing plaintiff's claims came instead from the doctrine of "preemption." What this meant, the Supreme Court explained, is where Congress has legislated in the field (e.g., laws regarding air pollution from power plants), there can be no federal nuisance law borrowed from classic principles of case law. In such instances the case law is trumped and regulatory authority and deference to it controls strictly in accord with statue. Very different from the *Massachusetts* holding, the high court did not seem to mind whatsoever in the *AEP case*; there might be numerous agency capture issues involved in EPA's non- or underregulation of plants' GHG emissions. The Court appeared pleased that EPA would use rationalism, balanced and informed assessment of competing interests (in this case described as "environmental concerns" versus "economic concerns"—as if the two were mutually exclusive), and its scientific expertise to appropriately regulate, if at all, the nastiest cause of all GHG emissions, those from coal burning power plants. The Court repeatedly reiterated that judges are not climate scientists and that these matters are best left to the regulatory experts at EPA. Interestingly, the Court concluded state (not federal) law claims based in traditions of case law [and state police powers—including perhaps criminal law] might still be open to the plaintiffs, and because the Circuit Court had not decided this issue, the case was remanded for possible review of state based remedies.

Still at the intermediate appellate level (meaning the case might yet reach the Supreme Court), there have been two notable cases recently: *Village of Kivalina v. ExxonMobil Corp and 663 F. Supp. 2d 863* (*N.D. Cal* (2009)); and *Comer v. Murphy Oil, U.S.A.*, 607F.3d 1049 (5th Cir. 2009).

The claimants in *Kivalina* constituted a village in northwest Alaska, and the defendants were some of the largest Western oil and energy companies. The District Court dismissed the village's claim it was being destroyed by sea level rise attributable to global warming. The court said that the village's claims involved purely nonjusticiable political questions and that a traceable injury from the defendant's emissions to the village's injuries could not be shown. This case has been appealed to the intermediate court level and is still under review.

In *Comer* the claimants included a number of Mississippi Gulf Coast residents affected by Hurricane Katrina, and the Defendants included a long list of America's oil and energy sector giants, as well as some major manufacturing firms. The Mississippi residents alleged nuisance and negligence (including fraud) by the defendants in contributing to climate change, which they claimed led to Hurricane Katrina in 2005. In dismissing the complaint, the court again used the doctrines of lack of standing for the plaintiffs and held the case presented a nonjusticiable political question. Under a bizarre set of circumstances, the plaintiffs filed an appeal and the case was assigned to a three-judge intermediate appeals court panel, of which only two heard the argument. When one of the two remaining judges then recused himself, the panel lacked a quorum. The case was then rescheduled, and another three-judge panel issued a ruling reversing the trial court's dismissal and finding the plaintiffs had indeed stated a justiciable case, Comer v. Murphy Oil, Co., 585F.3d 855 (5th Cir. 2009). The industry defendants then petitioned for rehearing *en banc* (meaning the entire 16 member court might hear the case). The nine active circuit judges who were not recused by then, barely constituting a quorum, granted the rehearing *en banc* by a vote of six to three. Perhaps the most interesting sociological fact of this case was that so many of the judges had to recuse themselves due to their financial interests in the energy sector—a very common situation in the Fifth Circuit, which comprises Louisiana, Texas, and Mississippi. In April 2010, one of the remaining nine judges recused themself, leaving just 8 out of 16 judges able to participate. The remaining eight nondisqualified judges then voted, six to two, to dismiss due to lack of quorum, thereby reinstating the trial court's original dismissal.

Another important federal climate change case which seems to have run its course was *Green Mountain Chrysler v. Crombie*, 508F. Supp. 2d 295 (D. Vt. 2007). In this case, the auto industry sued alleging Vermont's GHG emission regulations of automobiles was preempted by the Energy Policy and Conservation Act (EPCA) and also intruded into the field of foreign affairs by conflicting with federal foreign policy—which if true would likely constitute a political question. The court agreed with neither proposition and found that such regulations were within state police powers, that EPCA (which dealt with fleet miles per gallon) did not deal specifically with carbon emissions, and that there was no evidence for adducing that Vermont's regulations intruded into foreign affairs (Mounteer 2009; Martel and Stelcen 2007).

The recurring issues in these cases seem to be (1) the role of Agency—particularly the EPA and state variations thereof, (2) the law and economics ideological jurisprudence of the regular courts, which believes climate change can be solved by market forces or nominal economic regulations, (3) scientific testimony pertaining to difficulties in proving defendant's conduct caused claimant's harm, and (4) the pervasiveness of the "political question doctrine." These issues are discussed below.

In the American system, too often the agencies in charge of the science are captured (Franz 2011; Seis 2001). Agency capture is a concept that speaks to lack of agency accountability and political resolve due to various systemic forces (such as revolving doors, shared closely held information, divided loyalties, outright corruption, selective enforcement, selective cost–benefit analysis, etc.) within the regulated community, such that agencies ultimately pursue the agendas of those they are supposed to regulate instead of the political will of democratic majorities. Indeed, captured agencies lay down serious obstacles to fair, open, and rational decision-making, and therefore ultimately cause little economic and legal disruption to regulated industries. This is what the claimants in the *Massachusetts* case alleged was occurring at EPA prior to their bringing the complaint. Because the agencies and the regulated community speak the same complex language, and the courts have been indifferent—often deferring to agency expertise and discretion (e.g., the *AEP* case)—the process of regulation comes to serve private interests that are highly damaging to the climate (Wood 2007). Agency capture in the area of environmental regulation is well documented (Franz 2011). For instance, with the George W. Bush administration key agency positions affecting climate change were given to industry lobbyists, if not the industrialists themselves. These revolving door policies strengthened the link between government and the private sector and helped to promote national energy rules, policies, and practices that undermined climate science and hurt the climate (Lynch et al. 2010). Given the well-documented abuses inherent in capture, the widely heralded *Massachusetts* case loses much of its luster.

The most widely used agency principle is cost–benefit analysis, which the high court spoke to briefly in the *AEP* case. Cost–benefit analysis is also the most widely critiqued agency approach in the sociological literature. The critiques focus on how agencies use industry created data and outdated free-market analogies to reach decisions that the costs of environmental safety are too high (Lynch et al. 2008). Even so, cost–benefit styled regulation has come to mean the epitome of rationality. The US market and society are forced to respond to this artificial rationality in ways that undermine human freedom and ecologies—over time making people less aware of how compromised they are by the technologies driving climate change (Irwin 2010).

A standard neoconservative approach to climate change is that it is either going to be solved by free markets or that it is a kind of market failure to be fixed with nominal regulation. This jurisprudence frequently reflects Hardin's, The Tragedy of the Commons (1968), where each member of a community sharing a common commodity, due to so-called "rational self interest," keeps marginally exploiting that commodity until, to the detriment of the public interest, it is depleted. The community depletes the shared commodity because each member's selfish interest blinds them to the fact that the fractional exploitation they engage in is multiplied into a destructive aggregate of externalities when all engage in the same behavior. This model is an acceptable way to conceptualize climate change, with the atmosphere serving as the commonly held commodity, and GHG emissions constituting the marginal conduct. The model fails from a rights perspective, however, since it assumes that all parties sharing the commons are of equal power and right. The truth is, not all parties in the commons are equal, and this is particularly true of corporate members of the commons who are more highly motivated and bring to bear far more sophisticated means to expropriate the good of the commons for their own selfish ends—things like public relations and disinformation campaigns, lobbying, corruption of government officials, funding of questionable scientific studies, commission of a number of other serious but unenforced crimes, and even the creation of normative conduct via "management of demand" (see also Chap. 4). There is clearly, then, an environmental justice aspect to the tragedy of the commons that utilitarian law and economics theory misses because it fails to take into account imbalances in the distribution of powers.

A less critical understanding of the tragedy of the commons allows corporations to portray themselves as "redeemed sinners" offering consumers a choice in climate change reducing consumer products. The new mythical narrative of corporate America—the green company facilitating consumer choices that help, not hurt the environment is the latest "free market" conceptualization. If we all consume the right light bulb, and drive the right vehicle, everything will be ok, and our collective guilt assuaged (Sinden 2007), even though the underlying evil of the machinery of production, consumption, and growth demand models are left intact. The calculations of climate change risks are fit neatly into actuarial tables quantifying and managing risk (Irwin 2010). That the insurance industry has already started responding to the threat of climate change (e.g., regarding crop losses, building permits, fire responses, rate adjustments) in ways that undermine the so-called "free markets" approach put forth as rationalizations by the energy industry defendants, administrative agencies and too many judges have been lost on American society (Vanderheiden 2008). The legal risk assessment approaches, so far, make certain forms of humanity (the Inuits, the Island of Tuvalu) and animal life (the pica, the polar bear) expendable (Irwin 2010; Sinden 2007) and, the American courts, bound by rational choice economic models find themselves unable to describe such losses in meaningful terms (Skocz 2009).

The science of climate change has received significant attention in the legal literature. The two areas most addressed are (1) the science used to regulate and (2) the science used in the regular courts to prove harm (given that footnote 2 in the *AEP* case seems to have once again raised the specter the Supreme Court is staffed by a number of climate change deniers). The first issue is fairly easy. It is widely recognized by forensic and legal experts that the EPA, when it does use them, uses fairly good climate models. Here, it is industry that usually attacks—doing so on its "choice of model," or on the "application of a model in a clearly inappropriate setting," rather than on the accuracy of the models. Challenges to EPA models have usually failed, even as the models used are not as precise as those used in the regular courts for litigating causation and damages (Farber 2008). The second issue, of using climate change science to prove harm, is more problematic and a serious obstacle to climate change litigation in the USA. In order to prevail in a case, a claimant must show proximate cause, which means that they must show that it was substantially the defendant's conduct that caused their injuries. Recalling *Kivalina, AEP,* and *Comer*, there were serious questions as to whether plaintiffs could show proximate cause. This is because climate change is a global phenomenon. The worst defendant in the *AEP* case is the worst GHG polluter on earth, yet they contribute only 2.5% to all global warming worldwide. This would be a small percentage on which to hang substantial liability for causation of harms in a negligence or nuisance suit—though, of course, plaintiffs were merely seeking an injunction. An earlier *Comer* court called the prospect of this proof daunting, even in light of the low threshold; a "preponderance of the evidence" standard. Given the current state of evidence rules for expert witnesses under *Daubert v. Merrell Dow*, 509 U.S. 579 (1993) and its progeny, judges will, if plaintiffs can get past other standing issues, tend to reject the proffers of evidence of causation of harm (not the same as "causation of climate change") as lacking testability, peer review, known error rates, and general acceptance in the scientific community—Federal Rules of Evidence 702 and 703. It is likely judges will persist in exercising their role as gatekeepers of evidence, and that experts who would testify to specific causations of climate change causing harm to specific individuals or communities will not be permitted to go before juries. Even as the Government Accounting Office, the insurance industry or even oil and auto executives clearly indicate they believe in climate change harms, this is not sufficient for the courts to isolate specific enough harms from which to calculate damages caused by the phenomenon—the sheer numbers of polluters is a daunting fact to the American courts. However, as climate change is increasingly accepted as a scientific and social fact, one can expect standards of proof of harm may tend to lighten, and the courts will find a way to look to aggregate harms (Reeves 2009; Hsu 2008).

The "political question doctrine" is a much simpler issue to deal with, though it is a great obstacle in climate change litigation. Whether or not a judge believes a certain matter is one of politics (for the executive or legislative branches) or one that is ripe for judicial treatment is largely a personal preference. Clearly some hard and fast cases exist where the standard of what constitutes a political question is clear. In the area of climate change, however, judges opposing standing focus on the vaguer notion of "the alleged impossibility of a court to render an independent decision without offending another branch of government." Observation in these types of cases shows political question doctrine determinations are clearly an area where the judge's own ideology controls as to what he or she believes is an issue that it is prudential for the courts to hear and adjudicate and what is not (Tushnett 2002).

Non-American and International Courts

There is much stronger movement toward legal remedies concerning climate change occurring around the world than is occurring in the USA. There are international and regional agreements, protocols, and conventions now in play (White 2011). In many nations environmental protections that could be brought to bear on the problem of climate change have been imbedded into constitutional rights, or found in the jurisprudence of natural law.

According to Vanderheiden (2008) most classic and contemporary justice perspectives are simply inadequate to deal with the cosmopolitan and radical egalitarian issues climate change raises. Those impacted by climate change have brought actions in a wide range of jurisdictions (Osofsky 2006). The concept of a sovereign's obligation to those outside its political boundaries remains largely a foreign concept where, with climate change, international approaches need a focus on cross-border remedial actions, bio-diversity issues, the shared use of oceans and other resources. The conventions that have attempted to undertake this task are cast in high sounding and often enough inefficacious language. These conventions, conferences and reports include the 1948 United Nations (UN) Declaration of Human Rights; the International Covenant on Economic, Social and Cultural Rights (1966); the UN Conference on the Human Environment—the Stockholm Declaration of 1972 (Arts. 21 and 22); the Hague Declaration of 1989 (Art. 1) (also establishing the International Court of Justice [ICJ]); the 1992 UN Framework Convention on Climate Change (known as both UNFCCC, or the Rio Declaration of 1992); the 1994 Final Report of the Special Rapporteur, UN Commission on Human Rights 1994; the 1994 Draft UN Declaration on the Rights of Indigenous Peoples; the Kyoto Protocols of 1997; and Copenhagen 2009. With these and a number of other projects the international community has increasingly recognized an interconnected right amongst nations to an adequate environment supportive of life, dignity and well being that is not exploited in ecologically damaging ways by other nations, and which requires a future of cooperation for development of liability and compensation laws for victims of environmental damage caused by activities occurring beyond present jurisdictional bounds (Lynch et al. 2010; Vanderheiden 2008; Berat 1993).

For instance, although the nonbinding Rio Declaration has had difficulty in practice (Lynch et al. 2010), it spawned a number of concepts, including "intergenerational equity"; "sustainable development"; and "the precautionary principle." The 1994 Human Rights and the Environment, Final Report of the Special Rapporteur, UN Commission on Human Rights, and the 1994 Draft UN Declaration on the Rights of Indigenous Peoples, though both nonbinding, can together be read to mean there is an emerging "customary law" that inundation of coastal waters for certain peoples is an international human rights crime subject to ICC jurisdiction as well as universal jurisdiction in other nations (Sharp 1999). International covenants and accords are not useless exercises. They can help to establish "soft law," and *jus cogens*, which can then be treated by individual nations as enforceable international customary law in domestic courts. According to modern theory, a principle becomes part of customary international law when it is (1) widely adhered to by a number of states, and is acquiesced in by others, and (2) engaged in out of a sense of obligation. If the practice is uniform, the period of time it takes to become an international custom need not take long (Berat 1993). An increasing number of other documents reflect a similar sentiment toward customization of such soft laws regarding the environment and human rights. These include, inter alia: the 1981 African Charter on Human and People's Rights; Article 11 of the 1988 Additional Protocol to the American Convention of Human Rights; the 1988 Brazilian Constitution; the 1990 Namibian Constitution; Czechoslovakia's 1991 chartering of a right to live in a favorable environment and a right to have timely and complete information about the state of the environment; and the 1996 South African Constitution. The 1988 Brazilian Constitution, for instance, proclaims that all are entitled to an ecologically balanced environment essential for a healthy life and the state has the duty to protect and preserve the environment for present and future generations. Recent legal developments also include legal decisions in an increasing number of nations supportive of the reemerging environmental concept—the "public trust doctrine," which requires courts to actively hold government accountable to manage and protect nature for the benefit of the public. For instance, in the Philippines, the public trust doctrine was invoked in the case of *Oposa v. Factoran* (1993) to halt carbon intensive rain forest logging (Wood 2007); the doctrine was also used in India, in the case, *M.C. Mehta v. Kamal Nath* (India 1996–1997) and it is showing vigorous signs of life elsewhere, such as Pakistan, Uganda, Kenya, Nigeria, Ecuador, and Canada (Blumm and Guthrie 2012). As yet, the public trust doctrine has not been strongly tied to GHG emissions cases, but this is an area to watch closely. The public trust doctrine clearly raises global issues of environmental justice, since the first human communities to be affected by climate change seem to be those contributing least to the problem (Lynch et al. 2010).

Unfortunately, the world's per capita leader in GHG emissions—the USA, has resisted many of these international agreements related to climate change (e.g., Kyoto 1997—the USA citing concerns that China and India were to be exempted from GHG reduction processes). US courts have also refused to consider legal access along soft law or customary law lines, *Beanal v. Freeport-McMoran, Inc.*, 197F.3d 161 (5th Cir. 1999) (dismissing a case involving torture, murder and corruption, while finding "environmental responsibility" to be a term devoid of discernibly adjudicable standards). The *Massachusetts* case, and even the *AEP* case (though vaguely) both referenced the "unfair" or "unworkable" fact China and India were exempt from international agreements as a problem or potential problem for standing in proving necessary causation of harms. Interestingly, Vanderheiden (2008) has suggested elites, in fact, wanted India and China and other developing countries out of the pledge in order that climate harming industries from the USA might more easily be moved to these locations.

In the European Union (EU), much stronger efforts have been taken to reduce carbon emissions than in the USA. The EU is seeking a 60–80% reduction in GHGs by 2050 (Wood 2007). The UN Economic Commission for Europe Aarhus Convention of 1998 (Aarhus), based on the Rio Declaration is the seminal document for the courts and claimants to ensure this occurs in the EU (as well as a number of Central Asian countries who have joined the convention) (Stanley-Jones 2011). Aarhus seeks to protect the rights of all persons, and future generations to health and well-being. Accordingly, Aarhus guarantees to all EU citizens (and in some cases noncitizens) (1) access to information on the environment as a right, (2) public participation in environmental decisions as a right, and (3) access to justice on environmental matters as a right. Aarhus provides every citizen of the EU the opportunity to learn fully about and to seek to enjoin through the courts and without cost, any proposed action at any level of government or government related activity, that might possibly degrade the environment (Scannell 2010; Vanderheiden 2008; Kravchenko 2007).

Once in the courts, the inquisitorial legal system of the EU is better able to understand the relationship of scientific inquiry to law than is the US' adversarial legal system. Instead of punting scientific questions to captured regulatory agencies or leaving them to unequally financed parties motivated to produce science supportive of their claims or defenses, the EU legal system treats science as more unified (Pease 2011). There is less bias and conflict of interest in the science presented in EU courts, which must give reasons for departing from standard scientific findings such as those contained in IPCC reports. Here the "precautionary principle" works well, with the so-called Seveso Directives of 1982, 1996 and 2003 mandating all technology that might impact the environment or human health, flora, fauna, soil, water, climate, landscape, material assets, and cultural heritage must be tested rigorously with EIAs before release on the market. Member states may provide a single procedure consistent with Aarhus in order to fulfill the requirements of these directives. The "precautionary principle" is a rational and scientific principle that can be used for litigation under Aarhus, particularly at the agency level. The precautionary principle owes its existence to the Rio Declaration of 1992. There Principle 15 stated,

> [i]n order to protect the environment, the precautionary approach shall be... applied by states according to their capabilities. Where there are threats of serious or irreversible damage lack of full scientific certainty shall not be... reason for postponing cost-effective measures to prevent environmental degradation.

The ICJ also contains a precautionary principle at Article 3, which deals directly with climate change, and reads similarly. Several points are noteworthy, however, in showing the precautionary principle is susceptible to capture. As a practical matter, no climate change cases under the precautionary principle through the auspices of either Aarhus (via Seveso) or the ICJ have been brought (Kravchenko 2008). Although a number of scholars have looked at the issue, the ICJ is widely perceived as moving too cumbersomely for such an action (Berat 1993), and as yet Aarhus has only dealt with analogous localized catastrophic environmental situations (Kravchenko 2008). The EU and signatory states are only required to use the principle according to their capabilities and its use is largely nonbinding. The threats need to be considered "serious or irreversible"—yet no real definition is given as to what constitutes this standard. Presumably what agencies might come to call "less serious potential harms" will not merit application of the principle. The inclusion of "cost-effective measures" in the context of the precautionary principle leaves it open to interpretation by those holding the imbalance of power. Science tends to be at its best when it stretches the given norms of human perception, but constrained by concepts such as "lesser potential harms" or "cost–benefit analysis" is to guarantee a mundane science—one that will fail to overcome the tension between evidence and explanation so needed in this area (Irwin 2010). Even under the precautionary principle, a science is thereby produced that fails to comprehend or glosses over the externalities of climate change inherently felt in the lived experience of human, animal and plant life at the margins of existence (Skocz 2009).

One possible opening for climate change litigation under Aarhus in the EU is through EIAs. Unlike EIAs in the USA, EU EIAs must show concern for effects on the environment, not just on the local neighborhood. This is due to the EU commitment to carbon trading, which has not been well accepted in the USA. In this way, climate change cases in EU could occur at the very local level. It is not uncommon for a local land development that might cause flooding due to climate change to be stopped, or for major new building projects to not be approved if they will require additional parking spaces—cars being a major source of GHG emissions (Scannell 2010).

The EU took the Kyoto Protocol and the associated IPCC mandated emissions reductions very seriously (Transparency International 2011). The distinction between rich and poor countries and difficult GHG reduction targets were not seen as problematic and not seriously debated as in the USA. While post-Copenhagen, 2009, US credibility on climate change has fallen throughout the world, in the EU, it was understood China was a world leader in GHG emissions and India might soon be, but their per capita rates were and would remain well below many other nations. The question was asked in the EU: are we asking developing countries to pay for the damage we have historically caused by our luxury emissions and then, despite our greater capabilities to reduce current emissions, limiting their ability to get to our level or even one allowing bare survival emissions? (Vanderheiden 2008; Scannell 2010). Every EU state received a target for carbon emissions and other gases, with Europe on the whole accepting an 8% reduction (the suggested US reduction was 12%). Instead of balking at Kyoto, a Clean Development Mandate between rich and poor nations was vigorously implemented by Directive 2003/87, covering 11,000 heavy energy use installations and forcing large reductions in three phases.

Unlike the USA, in the EU it is seen that good laws, regulations, and wise mandates push innovation and sound business models. This has been demonstrated in oil and gas exploration and development (Weaver 2011), alternative fuels, carbon labeling, and the creation of a whole new industry of engineering firms to verify developed and developing nations compliance with emission mandates. There are numerous examples of climate legislation, such as Directive 2009/29 (The Renewable Energy Directive); Directive 2009/13 (The Carbon Storage Directive); Directive 2008/28 (Automobile Emissions Directive). These Directives have been used to address emissions trading, agricultural emissions, transportation related emissions, land use practices effecting emissions, or the funding for research and development of sustainable biofuels. As for vehicle emissions, the much more rigorous European standard is not expressed in kilometers per liter, but the more important "carbon emission per kilometer" (e.g., what's coming out of the tailpipe). Although EU cars are already more efficient, the long term goal reductions make US goals seem microscopic (see Chap. 12). If respective EU nations violate these directives and mandates they face suit in the European Court of Justice seeking to enjoin their practices (Scannell 2010).

In short, the litigation of climate issues in the EU has taken a different route due to better legislation and more impartial agency administration of climate change science. One prominent EU case shows how these principles work in practice. The case was a 2003 request for information submitted against Hermes Euler HG, an export credit agency, which finances major GHG emissions producing projects in developing countries around the world. The NGOs petitioning the court claimed the German Environmental Information Act as transposed into German law through EU directives gave them the right to information about the extent to which Hermes provides political and economic risk insurance to projects producing GHGs. The Ministry of Economics and Labor refused the request for information in 2004, and the NGOs brought an administrative action involving the Ministry and the financier in Berlin. The case ended in a quasi-settlement-order, whereby Hermes was required to provide detailed information on GHG implications of its energy production projects (Osofsky 2006).

In Australia, the case of *Australian Conservation Fund v. Latrobe City Council* (Victoria 2004) is noteworthy. There the Hazelwood Mine and Power Station (a U.K. owned subsidiary), which were burning local coal, were sued under state and national law to prevent global harms. The suit challenged limitations imposed on a state-appointed panel inquiry into environmental effects occurring under Australian law. The mine and power station, which provided over 20% of Victoria's base load electricity, was running out of coal and sought an additional coal field in the city of Latrobe that would keep the plant running through 2030. The Minister of Planning approved the panel inquiry's finding as to its EIA, but excluded climate change from those findings, triggering the NGO suit. Although the expansion ultimately went forward, the decision was hailed in certain sectors as an instance of the judiciary forcing an agency to behave in a certain way toward industry, thereby producing the first Victorian GHG reduction deed, establishing emission caps, providing for surrender of coal, setting milestones for reporting requirements, and encouraging the development of alternative energy sources. Greenpeace remained critical of the decision, claiming the stipulations were "window dressing" that permitted the plant to emit vast amounts of GHGs (Osofsky 2006). The Hazelwood Plant, said to be Australia's largest GHG emitter, continues to pose major climate change problems (Wilkinson 2009).

In the North and South Americas, the American Conventions on Human Rights of 1987 and 1988 promised to ensure the right of all to live in a healthy environment, requiring signatory states to promote the protection, preservation and improvement of the environment. The Inter-American Court of Human Rights (IACHR) was also established. So far, environmental activists have not been successful in bringing cases through this court and under these conventions (Berat 1993). For instance, the Inuit people of the Arctic filed a petition in 2005 with the IACHR claiming the acts and omissions of the USA in regard to climate change have violated human rights. The Inuits homes are melting into the permafrost, and their hunting sources for food are seriously threatened. The Inuits also complain of thinning ice that makes travel routes more dangerous (Osofsky 2006). In 2006, the IACHR dismissed the case, stating the information alleged is insufficient to determine whether human rights violations, as they are defined by the "American

Declaration" have occurred. The Inuit then requested a hearing, which was held in 2007 (Sinden 2007), apparently for informational purposes.

Another case of international law is important because it used both national constitutional law and (what was once thought to be a vague and unenforceable) international law (Berat 1993) to gain a victory in the courts. In the case of Gbemre v. Shell Petroleum Development Company (2005), citizens living in eight different Niger Delta communities won an injunction in Nigerian court in a lawsuit filed against Shell, the Nigerian National Petroleum Company and four other oil producing defendants and the attorney general of Nigeria. The complaint opposed the widespread practice of "gas flaring," which causes climate change as well as the release numerous toxic pollutants into the local ambient air (Sinden 2007; Osofsky 2006). The amount of natural gas burned off daily in Nigeria contributed more GHGs than all of sub-Saharan Africa combined. Gbemre and others filed a petition under Nigerian constitutional law, which protects fundamental rights to life and dignity, and also under several articles of the 1981 African Charter on Human and Peoples' Rights (the Banjul Charter), which affirms that all people shall have the right to a satisfactory environment favorable to their development. After filing the suit, the head of the NGO who had assisted the plaintiffs was detained by Nigerian authorities for interrogation. Not only did the court decide for Gbemre, forcing Shell to come up with a detailed plan to stop the flaring, but it also ordered the legislature to begin amending a statute pertaining to the case.

A final possible avenue for vindicating climate change harms might be the International Criminal Court (ICC) (Sharp 1999). The ICC was, by its founding document—the Rome Statute of 1998, intended to address traditional human rights abuses. Since that time, there has been a convergence in thinking on the relationship of environmental harms and human rights. While the Rome Statute provides some environmental protection, mapping this convergence under any number of approaches remains problematic (Geer 1998). The USA, for instance, refused to sign the treaty, citing concerns the ICC would enact new and "unacceptable" crimes of "universal jurisdiction" that could reach nonsignatory nations (Sharp 1999). It seems the Rome Statute does authorize the ICC to develop core crimes beyond the ones initially listed, including possibly ecocide, and "geocide" (the killing of the earth) (Berat 1993). There is also an emerging crime of "cultural genocide" (Geer 1998), which could be implemented via climate change. New conventions of environmental crime could be used to create "soft law" where crimes such as climate change human rights harms might rise to the level of customary international law—*jus cogens* crimes (Sharp 1999). *Jus cogens is* an international law principle referring to a higher law that holds together the rule of law, and whose nonobservance is itself an international crime since this may spell the demise of rule of law to an entire legal system. There have been a number of international environmental crimes enforceable for quite some time. These include mostly laws protecting animal species—fur seals (international criminal laws dating to 1911), whales (enforceable international criminal laws dating to 1931), fish (internationally criminal since 1953), polar bears (1976), and birds (1940). They have also included criminal laws against polluting oceans, particularly with oil (1958) and pollution from ships (1973) (Berat 1993).

Whatever approach they take, sovereignties ought to move toward a permanent and comprehensive international environmental legal system recognizing the unique and fragile nature of the environment as well as emerging human rights and the crimes that may be committed against them that presently lay outside the traditional reach and scope of the state (Berat 1993). Berat believes that other crimes traditionally handled at the international level (slavery, genocide, wrongful mass expropriations and displacements, aggressive wars, piracy) whether by international tribunals or by states exercising universal jurisdiction are, because they involve severe human rights degradations, consistent with harms to the environment like climate change.

In regard to issues of culpability and victimization, countries such as the USA are responsible for the devastating global situation, and non-OECD countries tend to be most at risk. The USA constitutes less than 5% of the world's population, but is responsible for 28% of current GHGs. There are issues regarding social control that are so broad criminology has not even dreamed of imagining them. Lynch et al. (2010) have called the George W. Bush administration's failure address or even recognize climate change, and its collusion with industry a crime against humans and nonhumans. While one may not expect Mr. Bush to be hailed into a US court to answer these charges anytime soon, it is only just that he (and a large number of elites like him) should have to worry about charges should they travel abroad. Mr. Bush's policies helped to victimize areas most vulnerable to climate change caused drought, extinction, famine, disease, and sea level rise, i.e., the developing world—places where there are also less resources available with which to cope with the suffering (Sinden 2007). Culpability and victimization are clearly areas where criminology should have a voice—yet shamefully, the criminal justice community, like the legal community, has largely been silent regarding the definitive harm issue of the twenty-first century.

A major difficulty with using international criminal law against those who damage the climate has to do with issues of culpability. Berat (1993) calls for a regime where those criminally liable in international courts would be rulers, public officials, corporate elites, and corporations. Are the harms of climate change so great they equal something like crime, and therefore warrant such a process? Despite the fact climate change involves harmful and frequently unlawful behavior, is likely caused by immoral motives (Vanderheiden 2008), is an example of state–corporate crime (Lynch et al. 2010) implicates human rights violations on a widespread scale (Vanderheiden 2008; Sharp 1999; Berat 1993), and could be used quite efficaciously under traditional legal doctrines regarding state and federal police powers, no one in the legal literature of the USA is yet calling for criminalization of climate change harms. This lack of criminal discussion is mystifying given the popularity of the punitive and retributivist approach, "polluter pays," enjoys throughout much of the world (Vanderheiden 2008). Perhaps the lack of criminalization of climate change conduct is due in large part to the powerful positions the "persons" hold who commit or allow it and the relative lowly positions of its victims—"stratification," or because of a given insular society's morphologies, or because of the distance between the perpetrator and victim? (Black 1976). Because a criminal problem may be difficult to control, committed diffusely, or involve situations where those committing it are seemingly above the law or have various motives and varying degrees of blameworthiness this is all the more reason for criminology to want to study it. Clearly, climate change harms, because they involve immorality and serious ethical lapses at the "edge of crime," are worthy of study as a sociological concept, if not as a legal one. In fact, the places where climate change harms acts occur have tended to be "criminogenic places" where less than ethical state and corporate actors come together (Lynch et al. 2010; Osofsky 2006). Transparency International (2011) has noted that on the global level a large number of very serious crimes are intertwined with or precursors to the processes that create climate change, including murder, corruption, torture, rape, fraud, mismanagement of resources, conflict of interest, embezzlement, theft, perjury, and many more. A harder theoretical situation persists, however, where an entire population might be criminally liable for harmful conduct—such as a nation of people who treat energy or transportation practices that are extravagant luxuries as if they were a necessity for survival (Vanderheiden 2008). It must be considered that if elites are brought into these kinds of cases they will effectively continue to claim the defense of "necessity" so long as the democratic masses in their nations continue to consume and be employed by the production of commodities harmful to the climate. In these sorts of defenses, the claim that the matter criminalized is really just a domestic policy decision controlled by market demands and subject to the traditions of state sovereignty and not international crime, will likely be strongly persuasive within the nation state itself, and lead to further resistance among such nations to seek avoidance of the jurisdiction of such international courts (Berat 1993).

We may conclude that climate change in the courts is now a "social fact" and that law will act along rules of criminological analysis (Black 1976). Green criminology, despite often being policy directed and undertheorized, is increasingly seen as a valid and important addition to the social research agenda of behavioral science programs. Such work includes environmental justice, feminist theory, risk society, organized crime, corporate–state crime, and white collar crime, among others. The central difficulty criminology has had dealing with environmental legal studies is that many legal environmentalists tend to discount the human role in the study of Nature, whereas criminology as a behavioral science is, by definition, always concerned with the role of society. Criminology struggles with the idea natural phenomena might have their own intrinsic value and ought to be protected for their own sake (Smith 2001). To adapt to the harms of climate change and better speak to the law, criminology will need to broaden its theoretical scope to consider, more deeply, the source of all law, which Nature is intent not to let us forget.

References

Berat, L. (1993). Defending the right to a healthy environment: Towards a crime of genocide in international law. *Boston University International Law Journal, 11*, 327.

Black, D. (1976). *The behavior of law*. London: Howard House, Emerald Press.

Blumm, M., & Guthrie, R. D. (2012). Internationalizing the public trust doctrine: Natural law and constitutional and statutory approaches to fulfilling the Saxion vision. *University of California Davis Law Review* vol. 45.

Farber, D. (2008). Modeling climate change and its impacts: Law, policy and science. *Texas Law Review, 86*, 1655.
Franz, A. (2011). Crimes against water: The non-enforcement of state water pollution laws. *Crime, Law and Social Change, 56*, 27.
Geer, M. A. (1998). Foreigners in their own land: Cultural land and transnational corporations—emergent international rights and wrongs. *Vanderbilt Journal of International Law, 38*, 331.
Gerrard, M. B. (2007). *Global climate change and U.S. law*. Washington, D.C.: ABA Publishing.
Hardin, G. (1968). The tragedy of the commons. *Science, 162*, 1243.
Hsu, S.-L. (2008). A realistic evaluation of climate change litigation through the lens of a hypothetical lawsuit. *University of Colorado Law Review, 79*, 101.
International Panel on Climate Change (IPCC). (2007, February 5). Climate change 2007: The physical science basis. *Summary for Policy Makers* 2.
Irwin, R. (2010). Climate change and Heidegger's philosophy of science. *Essays Philos, 11*, 16.
Kravchenko, S. (2007). The Aarhus Convention and innovations in compliance with multilateral environmental agreements. *Colorado Journal of International Environmental Law and Policy, 18*(1), 1.
Kravchenko, S. (2008). Right to carbon or right to life: Human rights approaches to climate change. *Vermont Journal of Environmental Law, 9*, 513.
Lynch, M. J., Burns, R. G., & Stretesky, P. (2008). *Environmental law, crime, and justice*. New York: LFB Scholarly.
Lynch, M. J., Burns, R. G., Stretesky, P. B., & Burns. (2010). Global warming and state-corporate crime: The politicization of global warming under the Bush administration. *Crime Law and Social Change, 54*, 213.
Martel, J., & Stelcen, K. (2007). Clean air regulation. In M. B. Gerrard (Ed.), *Global climate change and U.S. law*. Washington, DC: ABA Publishing.
McKibben, B. (1989). *The end of nature*. New York: Random House.
Mounteer, T. (2009). *Climate change deskbook*. Washington, D.C.: ELI Press.
Osofsky, H. M. (2006). The geography of climate change litigation: Implications for transnational regulatory governance. *Washington University Law Quarterly, 83*, 1789.
Pease, C. M. (2011). Science and the law: An international perspective. *The Environmental Forum, 28*(1), 18.
Reeves, C. R. (2009). Climate change on trial: Making the case for causation. *American Journal of Trial Advocacy, 32*, 495.
Scannell, Y. (2010). *Lecture Notes from July–August course on EU Community Environmental Law*. South Royalton, VT: Vermont Law School.
Seis, M. (2001). Confronting the contradiction: Global capitalism and environmental health. In G. Howard & G. Newman (Eds.), *Varieties of comparative criminology*. Leiden: Brill.
Sharp, P. (1999). Prospects for environmental liability in the International Criminal Court. *Virginia Environmental Law Journal, 18*, 217.
Sinden, A. (2007). Climate change and human rights. *Journal of Land Resources & Environmental Law, 27*, 255.
Skocz, D. (2009). Environmental management in the 'Age of the World Picture'. In L. McWhorter & G. Stenstad (Eds.), *Heidegger and the earth: Essays in environmental philosophy* (2nd ed.). Toronto: University of Toronto Press.
Smith, M. (2001). The face of nature: Environmental ethics and the boundaries of contemporary social theory. *Current Sociology, 49*(1), 49.
Stanley-Jones, M. (2011). *The Aarhus Convention: A blueprint for inclusive and accountable climate governance? In: Global corruption report: Climate change*. London: Earthscan Publishing.
Transparency International. (2011). *Global corruption report: Climate change*. London: Earthscan Publishing.
Tushnett, M. (2002). Law and prudence in the law of justiciability: The transformation and disappearance of the political question doctrine. *North Carolina Law Review, 80*, 1203.

Vanderheiden, S. (2008). *Atmospheric justice: A political theory of climate change*. New York: Oxford University Press.
Weaver, J. (2011). *Lecture notes from July-August Course on Oil and Gas: Energy Law*. South Royalton, VT: Vermont Law School.
White, R. (2011). *Transnational environmental crime: Toward an eco-global criminology*. London: Routledge.
Wilkinson, M. (2009, November 6). "Brown down" in Australia. *Center for Public Integrity*.
Wood, M. C. (2007). Nature's trust: A legal, political and moral frame for global warming. *Boston College Environmental Affairs Law Review, 34*, 577–603.

Cases Cited

American Electric Power Co., Inc. v. Connecticut, Slip Opinion No. 10-174, 20 June 2011
Australian Conservation Fund v. Latrobe City Council, (Victoria, 2004)
Beanal v. Freeport-McMoran, Inc., 197 F.3d 161 (5th Cir. 1999)
Comer v. Murphy Oil, Co., 585 F.3d 855 (5th Cir.2009)
Comer v. Murphy Oil, U.S.A., 607 F.3d 1049 (5th Cir. 2009)
Daubert v. Merrell Dow, 509 U.S. 579 (1993)
Green Mountain Chrysler v. Crombie, 508 F. Supp. 2d 295 (D. Vt. 2007)
Massachusetts v. Environmental Protection Agency, 549 U.S. 497 (2007)
M.C. Mehta v. Kamal Nath, (India, 1996–97)
Northwest Environmental Defense Center v. Owens Corning Corp., 434 F. Supp. 2d 957 (D. Or. 2006)
Oposa v. Factoran, (Phil. 1993)
Village of Kivalina v. ExxonMobil Corp., 663 F. Supp. 2d 863 (N.D. Cal. 2009)

Chapter 7
Environmental Enforcement Networks: Their Role in Climate Change Enforcement

Grant Pink and James Lehane

Introduction

> *If practitioners bite off too much, chances are they will choke. Bite of too little, and nobody will much care. Obviously an agency can take bigger bites than an individual or a department; and a consortium of institutions [like a network] can presumably take even bigger bites without being overwhelmed* (Sparrow 2008, p. 84)

Climate Change is a global issue which will impact upon society in numerous ways and at many levels. To a large extent most attention and discussion has focused on the scientific and political aspects of climate change. Speth and Haas go so far as to suggest that "[g]lobal climate change is the most threatening of the major global change issues. It is also the most complex and controversial" (Speth and Haas 2006, p. 23).

The nature, scope and diversity of the issues associated with climate change can be grouped generically into social, economic or environmental issues. Collectively, these issues are often referred to as "the triple bottom line" with the phrase most often used in the context of ecologically sustainable development or corporate social responsibility (Suggett and Goodsir 2002, p. viii; Elkington 1997). Given the interrelationship between these three issues, and the concerns of the general public regarding climate change, social and economic issues will briefly be considered, with the discussion predominantly focussed on the environmental aspects—particularly environmental regulatory responses.

The primary focus of this chapter will be on the compliance and enforcement capacity of government regulators and how they might benefit from networks to maximise their regulatory outcomes. Government regulators refers to agencies with

G. Pink (✉)
Latham, ACT, Australia
e-mail: grantpink@gmail.com

J. Lehane (✉)
Holt, ACT, Australia
e-mail: jimlehane@gmail.com

R. White (ed.), Climate Change from a Criminological Perspective,
DOI 10.1007/978-1-4614-3640-9_7,

regulatory responsibilities and includes those at all levels of government ranging from local through sub-national, national, regional and global.

Environmental crime and its associated law enforcement issues have been referred to as "wicked problems" (Australian Public Service Commission 2007; Rittel and Webber 1973, p. 161) and "wicked issues" (Shergold, cited in Briggs 2006, para. 10). Furthermore, Williams (2006) considers them to be problems that "... cannot be resolved by organisations and agencies acting autonomously, but rather needs concerted focus and action across all sectors" (Briggs 2006, p. 254). Given the complexity of environmental law enforcement, government regulators are increasingly working in various partnering relationships with industry groups, non-government organisations and more recently academics (Interpol, n.d., para. 8). These arrangements require government regulators to liaise with a range of stakeholders who have different (and sometimes diametrically opposed) and competing viewpoints and interests. Like all government agencies, government regulators are accountable to the general public and governments of the day. To assist with accountability and transparency government regulators must prioritise their efforts and budget accordingly. On the issue of budgets, the Organisation for Economic Cooperation and Development (OECD) note that:

> [t]hroughout the world, authorities responsible for enforcing environmental regulations and promoting compliance with environmental requirements are operating in the context of financial constraints. Such constraints can be a consequence of the general pressures on the state budgets or the changes in government policies, which may result in shifting the resources to address short-term priority problems away from environmental protection. Very often, environmental inspectorates are required to maintain, or even achieve higher performance with fewer resources (OECD 2005, p. 3).

In response, and in an effort to "do more with less", government regulators have increasingly sought to leverage off of on one another and other stakeholders to achieve efficiencies. Whilst environmental enforcement networks, to varying degrees, have assisted government regulators to achieve efficiencies—they should not be seen as a "cheaper option" or alternative to adequately resourcing and funding compliance and enforcement efforts. Environmental Enforcement Networks should instead be viewed in terms of their effectiveness, especially given that:

> ... better coordination, both between countries and between agencies within a country is essential to use resources more effectively, improve monitoring, and to facilitate enforcement (Royal Institute of International Affairs 2008, p. 3)

Several issues become clear when considering the government regulators engaged in climate change regulation. First, there are many of them. Second, they have access to different resources whether they are human, financial or technological, and their frequency of access to such resources could be variable. Third, resulting from the public sector reforms of recent decades, there are greater demands and expectations being placed upon governments to be more outwardly focussed (Tiernan and Althaus 2005, p. 5), outcomes based (Atkins 2005, p. 11) and operate more collaboratively (Winkworth 2006, p. 31). As a result, there is an understandable expectation that government regulators will need to cooperate with one another for mutual benefit to achieve efficiencies especially in circumstances where their interests intersect and overlap (Pink 2010a).

Cooperation between government regulators is considered to be a well-established practice (Scanlon 2010, para 9; Braithwaite 2006, p. 46; National Association of Attorneys General and United States Department of Justice 2003, p. 1). However, research conducted thus far (Pink 2010b; Farmer 2007, pp. 249–262; Taschereau and Bolger 2006) suggests that this cooperation is shaped by a number of factors such as whether or not the government regulator is:

- A mainstream law enforcement agency that is:
  - Engaged in more traditional law enforcement activities
  - Able and permitted to exchange information, intelligence and evidence for law enforcement purposes
  - Essentially self-contained in terms of its ability to access prosecutorial, scientific and technical services

or

- An established environmental enforcement agency that is:
  - Engaged in environmental regulatory activities
  - Able and permitted to exchange information, intelligence and evidence with co-regulators and mainstream law enforcement agencies
  - Able to pursue (either internally or within another government agency) a range of remedies or sanctions including those which are administrative, civil, and criminal in nature
  - Able to access (either internally or within another government agency) pre-eminently qualified scientific and technical services

This chapter will commence by considering the role that environmental enforcement networks (EENs) might play in relation to climate change enforcement. It will then reflect on how EENs have been utilised to coordinate enforcement efforts across other commodities and topics associated with environmental enforcement before moving to consider the establishment and growth of a regional environmental enforcement network (EEN). It will also consider a range of factors and issues which tend to shape EENs before concluding by outlining a limited set of capabilities that researchers consider are present in vibrant and effective networks (Taschereau and Bolger 2006, pp. 13–20). These attributes have been considered by the authors in the development of the Network Evaluation Matrix (NEM), which will also be introduced in full.

Background to the Issues

Environmental Enforcement

All regulators face a number of universal challenges, including the development and maintenance of institutional capacity to undertake enforcement activities (Pink 2008; Van der Schraaf 2008; Farmer 2007). The environmental enforcement arena is noted as having greater and distinct challenges inherent in its operations when compared with mainstream law enforcement (White 2008; Wijbenga et al. 2008).

With environmental protection legislation evolving over recent decades, components of these legislative regimes can be and remain relatively novel (Du Rées 2009; Bates 2006). The relative youth of the environment protection legislation tends to follow the proliferation of international agreements, conventions and treaties relating to the environment and environment protection. The Environmental Treaties and Resources Indicators (ENTRI) website contains in excess of 100 agreements commenced since 1989 (Socioeconomic Data Application Center, n.d.). This aspect is even more important given most countries have not yet introduced climate change legislation.

As a consequence, environment protection legislation does not have an extensive body of case law (Reeve 2002) or common law (Bates 2006) behind it, both of which assists in interpretation and application, and are considered critical enablers to effective enforcement. Given its recent introduction, climate change legislation will also largely operate during its early years without specific guiding legal precedents. This aspect however should be assisted given that a number of countries have recently introduced emissions trading scheme (ETS) legislation and associated regulations. New Zealand and the European Union are two such examples (The Harvard Project on Climate Agreements 2010, p. 10).

Climate Change Enforcement's Interrelationship with Other Crime Types

In respect to enforcement of environment protection legislation, it is not always a responsibility that falls to mainstream law enforcement agencies, such as the Police and Customs services. Nor does it naturally fall to existing Environmental Enforcement Agencies (EEAs), such as Environmental Protection Authorities.

Instead, due to factors such as machinery of government changes (including agency mergers and the establishment of new agencies), enforcement of environment protection legislation is increasingly being undertaken by emerging or less experienced EEAs. As a result, agency staff, who hitherto have performed administrative, policy and programmatic tasks, now find themselves having responsibility for a range of compliance and enforcement activities. This not only requires that they learn and perform these new activities—but they do so in a what is a more adversarial and litigious environment. This is consistent with the observations of Bricknell who states that "agencies who only recently adopted the mantle of regulator are still negotiating the regulatory culture" (Bricknell 2010, p. 114).

Macken (2011) commented on the need to strengthen the credibility of ETS legislation in the European Union following security and fraud related incidents. More specifically, Macken outlines a number of different problems that became evident in 2010 and 2011 including:

> one national Government's decision to resell Certified Emissions Reductions that had already been surrendered; Value Added Tax (VAT) fraud; phishing attacks on user pass-words; and hacking of various accounts including one National Registry (Macken 2011, p. 1)

These incidents resulted in the loss of five billion Euro to VAT fraud, and millions of Euro worth of permits respectively (Macken 2011, p. 3). Macken concludes by suggesting that notwithstanding the identified issues (and action already taken or entrain) that "some are already wondering whether money laundering will be the next!" (p. 6).

This provides clear indication that traditional crime can operate in conjunction with new environmental commodity areas, and are quick to adapt their practices and operations to new opportunities.

An added complication for Climate Change Enforcement (CCE) arises due to the fact that CCE will involve offences which span various crime types. Climate Change Enforcement will have aspects of

- Traditional crime (Moore 2012, p. 184; Epstein 1998, p. 145)
- Environmental crime (INTERPOL 2009a; UNODC 2008a)
- Crossover crime (UNODC 2008b)

Table 7.1 contains the various crime types, typical activities, and the lead and assisting agencies involved.

Environmental Enforcement Networks

The authors share an extensive combined experience, approaching two decades, with respect to environmental enforcement networks. This has included representing a regional Environmental Enforcement Network as officeholders at various subnational, national, regional and international conferences and events. Their experiences include the following:

- Conducting global research into EENs (Pink 2011, 2010b).
- Documenting the evolution of a regional EEN (Lehane and Pink 2011).
- Developing the Network Evaluation Matrix (NEM)—an assessment tool for EENs (Pink and Lehane 2011).

These aspects are now considered in turn.

Table 7.1 Crime types, typical activities, and the lead and assisting agencies

Crime type	Typical activity and example	Lead and *assisting agencies*
Traditional	• *Theft* of ETS permits, vouchers and credits	Police
	• *Fraudulent activities* associated with the above	*Customs, Environmental, Tax*
Environmental	• *Damage/Harm* to the environment caused by a failure to "offset" the impacts through the ETS permits, vouchers and credits	Environmental
	• *Non-compliance* with an ETS permit, voucher or credits	*Police*[a]
Crossover	• *Corruption*—bribing of government and non-government officials who influence issuance/trade of ETS permits, vouchers and credits	Police, Customs
	• *Money laundering*—of the funds and assets derived from the dishonesty	*Environmental, Tax*

[a]It should be noted that in some countries (e.g. Israel and the Netherlands for example) that environmental enforcement is undertaken by the police. See White (2008, pp. 198–199); Tomkins (2009, pp. 515–517)

Conducting Global Research into Environmental Enforcement Networks

Pink's (2010b) study assessed the utility of Environmental Enforcement Networks. The study considered 14 current and active EENs, at the sub-national, national, regional and global levels and involved eight research participants. The Environmental Enforcement Networks considered in Pink's study are listed alphabetically in Table 7.2.

The eight research participants held direct environmental compliance and enforcement responsibilities within national or sub-national Environmental Enforcement Agencies (EEAs). Research participants also had formal roles within one or more of the networks studied. The research participants had contributed, participated in and gained experience from:

- Seven countries
- Eight agencies
- 14 networks and
- Association with networks that had been established for between 6 and 20 years

The research data focussed on collecting information across four broad areas namely:

- *Involvement*
- *Value*
- *Effectiveness*
- *Support* for networks

Table 7.2 Networks considered in Pink (2010b, p. 23)

Abbreviation	Full Name	Type and Year Est.
AELERT	Australasian Environmental Law Enforcement Regulators network	Regional (2004)
CEC	Commission for Environmental Cooperation (North America)	Regional (1994)
CLAG	Combined Law Agency Group (New Zealand)	National (1999)
ENDWARE	European Network of Drinking Water Regulators	Regional (2005)
HEEPA	Heads of European Environment Protection Agencies	Regional (2003)
IACP	International Association of Chiefs of Police	Global (1989)
IMPEL	Implementation and Enforcement of Environmental Law (Europe)	Regional (1992)
INECE	International Network for Environmental Compliance and Enforcement	Global (1990)
Interpol ECC	Interpol Environmental Crimes Committee	Global (1992)
NEEP	Northeast Environmental Enforcement Project (North America)	Regional (1980)
NRIG	Natural Resources Investigations Group (Australia)	Sub-national (2001)
TEEN	The Environmental Enforcement Network (Ireland)	National (2004)
WEG	Wildlife Enforcement Group (New Zealand)	National (1993)
WSP	Western States Project (North America)	Regional (1986)

Source: Pink (2010b, p. 23)

The research data was initially sorted as part of an iterative thematic analysis. The results were then subjected to an analysis of Strengths, Weaknesses, Opportunities, and Threats (referred to as "SWOT"). Table 7.3 shows the key themes that emerged from an examination of each of the SWOT factors.

The study established that there is utility in networks across the countries and networks studied. In particular, the research highlighted that EENs had been used to great effect by environmental regulators across a broad spectrum of commodities and areas of environmental concern.

Additionally, the utility of EENs has been independently and further demonstrated across the following sample of environmental commodities and topics:

- Wildlife smuggling (Scanlon 2010, para. 8; ASEAN-WEN, n.d)
- Electronic waste and organised crime (INTERPOL 2009b)
- Ozone depleting substances (UNEP 2007)
- Capacity building—in terms of:
  - Technical and regulatory aspects of environmental legislation (IMPEL 2009, p. iii)
  - Joint and cross agency inspections at sea ports (Heiss et al. 2011; INECE-SESN 2010)

Table 7.3 SWOT Analytical Matrix—Environmental Enforcement Networks—key themes in terms of strengths, weaknesses, opportunities and threats

	Supportive	Detrimental
Internal (to networks)	*Strengths* • Enhanced contacts • Operational benefits • Accessing better practices	*Weaknesses* • Lack of active participation • Lack of resources • Criticality of network secretariats • Project completion rates
External (to networks)	*Opportunities* • Representation • Reporting • Benefits • Communication • Events • Knowledge management	*Threats* • Inability to sustain internal capacity • Loss of key staff • Inadequate information distribution

Source: Pink (2011, p. 5)

- Twinning partnerships—focussed on:
  - Environmental impact assessments (AECEN 2010)
  - Environmental regulation and improving industry environmental compliance (AELERT 2011b, n.p.)
- Training of environmental enforcement practitioners (O'Leary and Lynott 2011, p. 4; Pink 2008)

In fact, Speth and Haas go as far as suggesting that "global and regional networking is proving its importance daily" (Speth and Haas 2006, p. 138).

Documenting the Evolution of a Regional Environmental Enforcement Network

In November 2003, representatives from 12 environmental agencies from around Australia met to discuss the challenges they were confronting. During this meeting, it became clear that these agencies shared many common challenges, including:

- Confronting the cross-jurisdictional nature of environmental issues
- Improving the level of cooperation between agencies
- Promoting cultural change both internally and externally
- Reforming legislation to improve enforceability and ensure compliance
- Improving training standards and opportunities for staff
- Improving consistency and accountability in decision making (Australasian Environmental Law Enforcement and Regulators neTwork AELERT 2011a; Lehane and Pink 2011, p. 2)

The Australasian Environmental Law Enforcement and Regulators neTwork (AELERT) emerged as a network following its inaugural conference in

November 2004. Formally established in 2004, AELERT brought together environmental enforcement agencies with the aim of working together cooperatively (Lehane and Pink 2011, p. 2). AELERT experienced significant growth in terms of membership and areas of cooperation to be considered a substantial regional Environmental Enforcement Network (p. 1).

As a newly established network, AELERT was largely focussed on Australian based operational practitioners in the environmental compliance and enforcement arena. However, partner agencies from New Zealand joined AELERT during 2008, resulting in the national network expanding to one that was regional (Lehane and Pink 2011, p. 6).

During 2008, Bartel reported that AELERT was having an observable impact on member agencies and their operations:

> [a]gencies are developing "cultures" of professionalism, learning, and engagement, both with other agencies within and across jurisdictions, as well with those being regulated. This evolution of agency style means that agencies will be more effective in the analysis and development of regulatory practice as well as more effective "on the ground" to attain environmental goals, build and regain public trust, and provide public benefits (Bartel 2008, p. 441)

The initial efforts of AELERT were focussed on facilitating and coordinating national wildlife enforcement operations and holding an annual conference. Further, these efforts tended to be directed towards the practitioners within member agencies. Over time, as the activities and "reach" of AELERT grew it became increasingly important to ensure there was support from both senior managers within member agencies and relevant ministerial bodies. As a result, AELERT sought and gained approval to report to a Ministerial Council in 2008 (Lehane and Pink 2011, pp. 5–6). The result being that as a *bottom-up* network, AELERT has evolved to include *top-down* oversight and direction setting which is recognition that "AELERT does not operate in an operational, policy or network vacuum" (Lehane and Pink 2011, p. 8).

Active contributions by AELERT member agencies have resulted in a more integrated approach to environmental compliance and enforcement nationally and regionally. In fact, contributions have formed part of broader global environmental enforcement campaigns. The Australasian contribution to INTERPOL's global enforcement initiative "Operation RAMP" (focussing on wildlife crime) conducted in 2010, is a prime example of the levels of coordination and cooperation that has been realised through an environmental enforcement network (see INTERPOL 2010). The current period of fiscal tightening has enhanced the need for agencies to work together to realise joint benefits and outcomes (Lehane and Pink 2011, p. 7; OECD 2005, p. 3).

The ability of AELERT member agencies to work cooperatively and deliver numerous outcomes for both government and member agencies has been demonstrated time and time again. By delivering tangible benefits to contributing member agencies, AELERT is now well recognised in various international forums as a successful network model.

In documenting the evolution of AELERT it became apparent that the existence, deliverables, and future of the network are dependent upon a result from the efforts of a small number of key people within key agencies throughout the network. In essence, the network's future is neither self-sustaining nor guaranteed. This was also the case in the majority of EENs mentioned in the previous study (Pink 2010b).

As a result, the authors formed the view that, despite sharing relatively common purposes, environmental enforcement networks are each unique. Part of this is due to the varying circumstances under which they were established—this is a viewpoint shared by Farmer (2007) and Fleming and Wood (2006). A range of factors influence the development of EENs and ultimately determine their relative capability and potential maturity as a network. This thinking led to the development of the Network Evaluation Matrix which is now considered.

Developing the Network Evaluation Matrix: An Assessment Tool for EENs

The purpose of the Network Evaluation Matrix is to apply an assessment tool to categorise the maturity levels of existing networks and establish a typology of networks. The Network Evaluation Matrix is based upon:

- Five levels of network maturity
- Five core functional categories of networks

First, during the sixth International Network for Environmental Compliance and Enforcement (INECE) conference in 2002, discussions were had in relation to the levels of network maturity. Participants from six regions (South America, Africa, Europe, Central America, North America and Asia Pacific) gathered separately within their respective regional networks and considered a range of issues affecting networks. They determined that networks fell within the three categories: *absent, fragile* or *well established* (Jones 2002, p. 464).

Whilst *absent, fragile* and *well established* are reflective of several fixed levels of networks, these three levels are possibly limiting, as they do not capture and reflect the ongoing evolutionary processes of networks, as well as the potential for regression. In considering the earlier observations of the INECE model alongside the foundational Capability Maturity Model (CMM), which was an assessment method designed to consider business processes (Carnegie Mellon University 2011), the authors considered it useful to include *emerging* and *maturing* as two additional transitional phases. These additions result in five comprehensive and representative levels of network maturity. These phases are as follows:

- Absent
- Emerging
- Fragile
- Maturing
- Well established (Pink and Lehane 2011, p. 4)

Second, in relation to the five core functional categories of networks—the authors consider that five major themes present core criterion by which it is possible to assess an environmental enforcement network. These criteria are the following:

- Membership
- Finances
- Governance
- Support
- Deliverables (Pink and Lehane 2011, p. 5)

Each criterion is further refined with three sub-criteria, as shown on the Network Evaluation Matrix (NEM) reproduced at *Appendix*. The NEM readily shows paths for the evolution of EENs through various phases of development and maturity.

The Network Evaluation Matrix, as a new network evaluation tool, was presented during a Summit of Regional Networks during the 9th INECE Conference. The Network Evaluation Matrix was accepted by participating Environmental Enforcement Networks confirming its relevance across a number of both regional and thematic EENs. This acceptance was evidenced by the fact that the INECE Secretariat made a call for all regional EENs to consider the matrix categories themselves as a form of self-assessment. A compartmentalised version of the NEM was then published as part of the conference proceedings (Pink and Lehane 2011).

The Network Evaluation Matrix deliberately excludes direct reference to specific commodities and professional streams within the environmental regulatory spectrum. This was done to reduce the risk of any one area or criterion being given priority or additional weighting, and the fact that the concept of evaluating a network's operation goes beyond issues of commodity, topic or professions. Further, it is considered that the NEM has application across other networks irrespective of whether they involve government agencies, the private sector and/or not-for-profit sector.

Networking: Is It Appropriate for Climate Change Enforcement?

Networking, whilst effective is considered by some (White 2011b; O'Flynn 2008; Himmelman 2001, 2002) to be the most preliminary of four levels of *working together*, and the others being cooperation, coordination and collaboration, respectively. The differences in the forms of working together and the exchanges involved in the various relationships has been well explained by these authors and serves to strengthen the process of Environmental Enforcement Network maturity as seen in the Network Evaluation Matrix.

All forms of working together come at a cost to participating entities. Further, none should be seen as a panacea. On these aspects, O'Flynn highlights the fact that:

> It is also important to be realistic about the trade-off between effort and reward. As strategies for working together become more complex—that is, they move towards the collaborative

Table 7.4 Continuum of relationships

Networking	Coordinating	Cooperating	Collaborating
• Exchange of information for mutual benefit • Informal relationship • Minimal time and trust • No sharing of resources	• Exchanging information for mutual benefit • Alter activities • Formal relationships • Requires moderate time and trust • Minimal sharing of resources	• Exchanging information • Alter activities • Sharing resources to achieve a common purpose • Formal relationships • Substantial time and trust required • Share resources • Some sharing of risk and rewards	• Exchange information • Share resources • Enhance capacity of another to achieve a common purpose • Formal relationship • Extensive time and trust required • Share, risks, responsibilities and rewards

Adapted from Fig. 2.1: Continuum of relationships (Government of Victoria 2007, p. 5)

> end of the scale—investments and costs intensify. … As Huxham and Vangen (2004) explain, the potentiality of collaborative advantage must be weighed against the hard grind of genuine collaboration or, in their words, collaborative inertia (O'Flynn 2008, p. 189)

Table 7.4 details a *Continuum of Relationships* and shows the key elements of each phase of the continuum.

Is Networking Enough?

Only time will tell if networking is enough for an issue as critical as Climate Change Enforcement. However, doing nothing is not an option, climate change regulators will be seen as negligent if they do not give serious consideration to establishing a Climate Change Enforcement Network (CCEN).

In their deliberations, climate change regulators should take into consideration successes that Environmental Enforcement Networks have brought to a broad spectrum of commodities and areas of environmental concern. As White (2011a) suggests, they need to recognise that "networking provides a practicable basis for intervention in areas that are by their nature complex and multifaceted" (p. 138). Steiner also highlights that "global problems need global partnerships … [and] enforcement networking is one small example of the benefit of such cooperation" (Steiner 2007, p. 2).

The International Network for Environmental Compliance and Enforcement (INECE) as a "network of networks" (INECE, n.d) has assisted many regional and thematic Environmental Enforcement Networks in their establishment and ongoing activities (Gerardu and Zaelke 2005). Focusing on a specific commodity issue, the INECE Climate Compliance Network runs workshops and produces publications which support practitioners working in the field of climate change compliance and enforcement.

Future Directions

In relation to establishing a Climate Change Enforcement Network (CCEN), there are short-term and long-term activities worthy of consideration. The more immediate and short-term activities should focus on the practical aspects of establishing a CCEN. The long-term activities should focus on assessing the relative maturity of the CCEN and conducting further research to determine the network's usefulness to ensure it remains relevant to members.

Short-Term Activities

Entities involved in climate change enforcement should come together in a coordinated way to discuss and assess their preparedness, desire and need to unite their efforts to combat climate change related crime. Given the above, and with *collaboration* as the ultimate goal, it is pragmatically appropriate that *networking* will be the most likely point at which climate change regulators should commence their interactions with one another.

Whilst this chapter has focused on Environmental Enforcement Networks, Slaughter (2004) suggests that there are three types of networks that may be worth considering. They are *information networks*, *enforcement networks* and *harmonisation networks*. Slaughter suggests that these "networks have overlapping functions—harmonisation and enforcement networks also exchange information and offer assistance; information networks can also make common policy for their members under certain circumstances" (Slaughter 2004, p. 52).

In considering the different networks:

- *Information networks* are:
  ... the glue of any trans-governmental network is the exchange of information and ideas. ... [simply] [p]ut a group of environmental regulators ... in a room and they will begin talking about different techniques of regulation, commiserating about common problems, and brainstorming new approaches (Slaughter 2004, p. 52).
- *Enforcement networks*—primarily focus on:
  ...enhancing cooperation among national regulators to enforce existing national laws and rules. As the subjects they regulate ... move across borders, they must expand their regulatory reach by initiating contact with their foreign counterparts (Slaughter 2004, p. 55), and

  At a very concrete level, enforcement cooperation is exactly the sharing of information and the collaborative development of specific enforcement strategies in individual cases. The next step is cooperating in strategic priority setting and targeting... (Slaughter 2004, pp. 56–57).
- *Harmonisation networks*—operate on the basis that "... regulators may work together to harmonise regulatory standards, ... [w]ith the overt aim of achieving efficiency" (Slaughter 2004, p. 59).

There is much to consider in relation to *networking*. One of the most important considerations should relate to what an effective Climate Change Enforcement Network (CCEN) might involve and entail. This chapter contains a great deal of information to guide the establishment of a CCEN.

In a study relating to *Capacity, Change and Performance,* Taschereau and Bolger (2006) developed a paper on *Networks and Capacity.* They adopted a broad definition of networks as being:

- Groups of individuals and/or organisations
- With a shared concern or interest
- Who voluntarily contribute knowledge, experience and/or resources for shared learning
- Joint action and/or to achieve a shared purpose or goal
- Who rely on the network to support their own objectives (Taschereau and Bolger 2006, p. 3)

Taschereau and Bolger suggest that:

> Individuals and organisations come together through a network around a common purpose, … [especially] … if they see a potential for increasing the capacity to achieve that purpose, either through sharing of information or joint action (Taschereau and Bolger 2006, p. 7)

This is reported especially in circumstances where there is a "sense of urgency, sense of frustration, … [and] possibilities afforded by information and communications technology" (Taschereau and Bolger 2006, p. 7). These circumstances seem well aligned with what is known of the challenges associated with climate change enforcement.

More specifically, Taschereau and Bolger go on to detail a set of capabilities that they consider exist in vibrant and effective networks. They include the following:

- Informal leadership
- Legitimacy and Collective Identity
- Technical expertise and resources
- Facilitating participation
- Managing and serving the network
- Communications and management systems
- Adaptive capacity (Taschereau and Bolger 2006, pp. 13–20)

These capabilities appear to be of particular relevance to those considering establishing or participating in a Climate Change Enforcement Network. Moreover these capabilities are complimentary to the core components of the Network Evaluation Matrix, with each capability having phases suited to maturing over time.

When considering networking and networks, Taschereau and Bolger (2006) suggest that consideration should be given to *purpose*, *form*, and *nature of participation.* These are critical considerations when considered against what Rhodes (2006) calls the "sour laws of networks" when describing networks. Rhodes suggests that networks struggle to deal with problems associated with co-ordination, mixing government structures, and ownership. Such problems would reflect through changes in the maturity level within the Network Evaluation Matrix.

In his chapter on Networking, Farmer (2007) describes the structure and function of six more formally established networks of varying ages. He concludes with a checklist for environmental enforcement agencies that are contemplating participating in networks. The Checklist asks:

1. Has the environmental enforcement authority a designated person responsible for coordination with the network?
2. Is it responsible for part-funding of network activities and, if so, is this fully incorporated in relevant financial planning?
3. Are effective mechanisms in place to identify staff members who might most effectively participate in relevant network activities?
4. Are effective mechanisms in place to disseminate the results of network activities to those in the environmental enforcement authority who would benefit? (Farmer 2007, p. 262).

Perhaps, one of the most important considerations will be determining whether any such Climate Change Enforcement Network would be a stand-alone network or form part of an existing network. The final decision should be informed by a thorough and detailed analysis of the similarities and differences (information, enforcement, harmonisation, and policy) of existing networks operating in the climate change space. Other considerations should include duplication of effort, competition for the same or similar network resources and dilution of network benefits.

Long-Term Activities

In the longer term, two activities are recommended in relation to a Climate Change Enforcement Network. First, after establishment (which will most likely involve a period of substantial growth based upon the experiences of other networks) and consolidation an assessment utilising the Network Evaluation Matrix should be undertaken to determine the relative maturity of the CCEN. The benefit being that it would assist in "identifying areas in which to concentrate effort in order to advance to the next level of maturity" (Pink and Lehane 2011, p. 13).

Additionally, further research using a mixed method approach as part of an exploratory study should be undertaken. Such a study would examine how members have actually used the Climate Change Enforcement Network. Furthermore, it would also determine the network's perceived utility from the perspective of a small sample of members (ideally drawn from a cross section of the membership involved in operational, policy, and management roles). The benefit being that it would enable the network to capture the "wants" and "needs" of the membership so that it might modify its "services" and/or "products" so as to remain relevant to members.

Conclusion

Environmental Enforcement Networks provide unique utility in the particular field of environmental compliance and enforcement (Gerardu and Zaelke 2005). Further, networks have the ability to coalesce environmental compliance and enforcement expertise, which enables agencies to engage in bilateral or multilateral projects and initiatives (Pink 2011, p. 6).

Environmental Enforcement Networks also provide individuals, teams, and agencies with access to a vast array of environmental compliance and enforcement expertise. Practitioners and network office holders particularly value access to this expertise. As reliance on networks grows, both groups should work to continue network progress, especially in times when resources are being reduced or are subject to increased scrutiny (Pink 2011, p. 6).

The current period of fiscal tightening (resulting in reduced agency budgets) combined with the cross-jurisdictional and transnational nature of climate change enforcement only reinforces the need for agencies to work together within a Climate Change Enforcement Network. Such a network provides a framework for enhanced collaboration, on this issue, Fleming and Wood (2006) consider that:

> [n]ot only do networks provide the opportunity for different forms of knowledge and capacity to be integrated in the furtherance of shared outcomes; they also provide the opportunity for resources (material and human) to be leveraged (p. 4).

When considering globalisation and environmental harm, White suggests that "[t]he challenges are clear; environmental issues are diverse and the time for action is now" (White 2010, p. 17). Given the global nature and extent of the environmental harms being attributed to climate change, combined with developing social and economic harms, *now* is the time for climate change regulators to act.

Environmental Enforcement Networks provide climate change regulators with a tried and tested means to assist them with *how* to act—especially in building and enhancing their compliance and enforcement capacity.

Appendix: The Network Evaluation Matrix

Absent criteria				
Members	Finances	Governance	Support	Deliverables
N/A	N/A	N/A	N/A	N/A
Emerging criteria				
Members	Finances	Governance	Support	Deliverables
Membership	*Budget*	*Access*	*Liaison*	*Events*
Typically low within network's sphere of influence Restricted to core member agencies or individuals operating within network's sphere of influence	Nonexistent or modest to initiate/commence the network	Member access to foundational and guiding documents No open source access to networks guiding documents	Limited or no liaison with other networks	A few events for members Limited number of members working together on specific projects
Leadership	*Contributions*	*Review*	*Support base*	*Outcomes*
Few individuals/member agencies take lead role across network	Limited in-kind contributions restricted to core member agencies or individuals	Informal review by foundation members	Network effectiveness reliant on core members providing ad hoc coordination	Prospective members considering potential of network Limited functionality within network to deliver outcomes

(continued)

(continued)

Emerging criteria				
Members	Finances	Governance	Support	Deliverables
Value	*Project funding*	*Structures*	*Supporting functions*	*Products*
Negligible or questionable value to non-core members Core members only active participants with few benefits realised	Non-existent or minimal	General Aims and Objectives laid down Typically developing and fluid No written procedures	Base level administrative tasks undertaken for network A few key individuals maintain the network	Low number of publications disseminated Publication limited to contributions of members
Fragile criteria				
Members	Finances	Governance	Support	Deliverables
Membership	*Budget*	*Access*	*Liaison*	*Events*
Generally increasing across significant actors within network sphere of influence	Contributions are spasmodic and are made by small number of sources Not sustainable	Central repository of information and communiqués to members Limited open source access to network's guiding documents	Emergent and exploratory liaison with other networks undertaken Information sought from other networks	Irregular events with increasing frequency Preliminary attempts to undertake cooperative exercises
Leadership	*Contributions*	*Review*	*Support base*	*Outcomes*
Wider range of key individuals/member agencies taking on leadership roles across network	Few core member agencies/ individuals providing in-kind support	Regular changing of operating rules, Constitutional documents, etc. to fit circumstances	Growing level of support base Increasing numbers of interested individuals	Low level coordination to deliver outcomes Members seeking benefits

Value Tangible benefits available for active members Most members inactive/passive and question value of network Low level communications disseminated to members from core members	*Project funding* Very limited, if available Sourced from lead member in project team	*Structures* Guiding documentation developed reactively to situations. Clear Aims and Objects set out for network	*Supporting functions* Central support function to support membership and administrative tasks only	*Products* Infrequent publications being developed and disseminated Range of network members contributing to publications
Maturing Criteria				
Members	Finances	Governance	Support	Deliverables
Membership Expanded to reach critical/core number of potential members within relevant network area of operation	*Budget* Sufficient funding for network to continue for short to medium term (2–3 years)	*Access* Increasing access of members to network steering/guiding documents Open source access to limited network produced documents	*Liaison* Interest by and liaison with other networks Capacity and capability discussions to identify areas of commonality	*Events* Held on semi-regular basis Delivered across network Range of events relevant to all members Restricted range of members working conjointly
Leadership Critical mass of individuals/member agencies take on leadership roles Wide range of membership not represented in leadership roles	*Contributions* In-kind support provided by central member agencies/individuals	*Review* Third Party review of constitutional documents and guidance considered	*Support base* Widening support base across network sphere of influence	*Outcomes* Member agencies receiving identifiable benefits Individuals professionally realise benefits

(continued)

(continued)

Maturing Criteria

Members	Finances	Governance	Support	Deliverables
Value	*Project funding*	*Structures*	*Supporting functions*	*Products*
Leading members displaying demonstrable benefits through active participation Level of inactive/passive members reducing Increasing engagement and communication between members	Project funding—increasing but ad hoc Project leaders are able to source limited funding to support project	Proactive development of guiding documentation with consultation across membership Development of Operation and Strategic Plans to set direction of network	Administrative tasks major part of function Low level of coordination and project capacity	Are of useable standard and considered as a step toward better practice across industry Contributions from those outside of network being included

Well Established Criteria

Members	Finances	Governance	Support	Deliverables
Membership	*Budget*	*Access*	*Liaison*	*Events*
Maximum or near maximum of possible members within relevant operation of network coverage	Secured on permanent basis Sustainable arrangements	Open and transparent access to network steering papers across membership and support base Foundational and guiding documents available as open source	Level of support from other networks seen through interaction and joint activities Good working relationships between networks	Held regularly and well attended Coordinated centrally for delivery across membership Members readily work collaboratively

Leadership Large proportion of members taking leadership roles across network	*Contributions* High proportion of members contributing in-kind support to projects, events and initiatives	*Review* Governance structures reviewed and consistent with better practice Subject to external scrutiny and review	*Support base* Strong support and contribution base from practitioner and senior management alike	*Outcomes* Delivering tangible benefits to members Serve as attractor, drawing new members into network Members overtly promoting the benefits of involvement
Value Majority of members realise benefits of membership attained through active participation Open communication across members	*Project funding* Projects driven by/within network are readily funded Projects undertaken by network bring in associated funding	*Structures* Robust written governance structures in place Membership well represented on guiding body	*Supporting functions* Central function for coordinating network activities, project and events Administrative tasks are a minor part of the function	*Products* High standard and considered better practicc across industry Subject to review and improvement process Wide range of contributors to network publications

References

Asian Environmental Compliance and Enforcement Network (AECEN). (2010). *Nepal and Malaysia Finalize Twinning Partnership Work Program*. Website – News release, 5 May 2010. Retrieved 2011, July 18 from http://www.aecen.org/nepal-and-malaysia-finalize-twinning-partnership-work-program.

Association of Southeast Asian Nations Wildlife Enforcement Network (ASEAN-WEN). (n.d). *Wildlife Enforcement Network: Action Update July-September 2010*. Retrieved 2011, July 18 from http://www.asean-wen.org/index.php?option=com_docman&task=doc_download&gid=45&Itemid=96.

Atkins, P. (2005). Developing transformational leadership capability in the public service. *Public Administration Today*, *3*, 11–17.

Australasian Environmental Law Enforcement and Regulators neTwork (AELERT). (2011a). *Background*. Retrieved 2011, January 06 from http://www.aelert.com.au/?page_id=5.

Australasian Environmental Law Enforcement and Regulators neTwork (AELERT). (2011b). *Australian-Vietnamese Twinning Program—Phase 2*. Website – News Release, 24 March 2011. Retrieved 2011, July 18 from http://www.aelert.com.au/?p=1274.

Australian Public Service Commission. (2007). *Tackling wicked problems—A policy perspective*. Canberra: Commonwealth of Australia.

Bartel, R. (2008, April). *The formal and informal orders in land clearance regulation in Australia*. Paper presented in International Network for Environmental Compliance and Enforcement (INECE), 8th International Conference on Environmental Compliance and Enforcement Proceedings. Cape Town, South Africa (pp. 435–443). London: Cameron May.

Bates, G. (2006). *Environmental law in Australia* (6th ed.). Chatswood, New South Wales: LexisNexis Butterworths.

Braithwaite, V. (2006). *Ten things you need to know about regulation but never wanted to ask*. [Occasional Paper 10] Regulatory Institutions Network, Canberra: Australian National University.

Bricknell, S. (2010). Environmental crime in Australia. In *AIC Reports: Research and public policy series* 109. Canberra: Australian Institute of Criminology.

Briggs, L. (2006). *Project management and organisational change*, speech delivered at the Australian New Zealand School of Government - Prime Minister and Cabinet conference, 21 February 2006. Retrieved 2008, October 1 from http://www.apsc.gov.au/media/briggs 210206.htm.

Carnegie Mellon University. (2011). *Capability maturity model: Integration*. Retrieved 2011, March 14 from http://www.sei.cmu.edu/cmmi/.

Du Rées, H. (2009). Can criminal law protect the environment? In R. White (Ed.), *Environmental crime: A reader* (pp. 638–655). Devon: Willan Publishing.

Elkington, J. (1997). *Cannibals with Forks: The triple bottom line of 21st century business*. Oxford: Capstone.

Epstein, J. (1998). State and local environmental enforcement. In M. Clifford (Ed.), *Environmental crime: enforcement, policy and social responsibility*. Maryland: Aspen Publishers.

Farmer, A. (2007). *Handbook of environmental protection and enforcement: Principles and practice*. London: Earthscan.

Fleming, J., & Wood, J. D. (Eds.). (2006). *Fighting crime together: The challenges of policing and security networks*. Sydney: University of New South Wales Press Limited.

Gerardu, J. J. A., & Zaelke, D. (2005). The importance of international environmental enforcement networks. *Environmental Law Network International*, *2*, 3–7.

Government of Victoria. (2007). *Victorian approaches to joined up government: An overview*. Melbourne: Government of Victoria.

Heiss, R., Ruessink, H., Isarin, N., Koparova, M., & Grabiel, D. (2011). International hazardous waste inspection project at seaports: results and recommendations. In *INECE 9th International Conference on Environmental Compliance and Enforcement, Conference Proceedings, 20–24 June 2011, Whistler, British Columbia, Canada* (pp. 192–203). Washington: International Network for Environmental Compliance and Enforcement.

Himmelman, A. T. (2001). On coalition and the transformation of power relationships: Collaborative betterment and collaborative empowerment. *American Journal of Community Psychology, 29*(2), 277–284.

Himmelman, A. T. (2002). *Collaboration for a change: Definitions, decision-making models, roles, and collaboration process guide*. Minneapolis: Himmelman Consulting.

Huxham, C., & Vangen, S. (2004). Doing things collaboratively: realizing the advantage or succumbing to inertia? *Organizational Dynamics, 33*(3), 190–201.

Implementation and Enforcement of Environmental Law (IMPEL). (2009). *Enforcement of EU Waste Shipment Regulation: "Learning by doing"*. Interim Project Report, 12 October 2009. Retrieved 2011, January 18 from http://www.vrominspectie.nl/Images/Interim%20%20Report%20IMPEL%20TFS_tcm293-273297.pdf.

International Criminal Police Organization (INTERPOL). (2009a). *Environmental crime*. Retrieved 2011, September 14 from http://www.interpol.int/content/download/793/6403/version/14/file/PST03.pdf [Fact Sheet].

International Criminal Police Organization (INTERPOL). (2009b). *Electronic waste and organised crime assessing the links: Phase II report for the Interpol pollution crimes working group*. France: Interpol.

International Criminal Police Organization (INTERPOL). (2010). *INTERPOL co-ordinated operation targeting illegal trade in endangered reptiles leads to arrests and seizures worldwide* Retrieved 2011, December 14 from http://www.interpol.int/News-and-media/Photos [Media Release].

International Criminal Police Organization (INTERPOL). (n.d.). *The INTERPOL global complex for innovation*. Retrieved 2011, December 6 from http://www.interpol.int/About-INTERPOL/The-INTERPOL-Global-Complex-for-Innovation.

International Network for Environmental Compliance and Enforcement (INECE). (n.d.). *Raising awareness, developing networks, strengthening capacity.* [Brochure].

International Network for Environmental Compliance and Enforcement—Seaport Environmental Security Network (INECE-SESN). (2010). *International hazardous waste inspection project at seaports: results and recommendations*. Retrieved 2011, June 18 from http://www.inece.org/seaport/exercise/INECE_SeaportInspectionProjectOutcomes_22dec.pdf.

Jones, D. (2002). Summary of plenary session #9: Reports of regional meetings and workshops. In *INECE 6th International Conference on Environmental Compliance and Enforcement, April 15–19, 2002 San Jose, Costa Rica, Proceedings Volume 2*. (pp. 457–465). Washington: International Network for Environmental Compliance and Enforcement (INECE).

Lehane, J., & Pink, G. (2011). Evolution of a regional environmental enforcement network: The Australasian Environmental Law Enforcement and Regulators neTwork (AELERT). In *INECE 9th International Conference on Environmental Compliance and Enforcement, Conference Proceedings, 20–24 June 2011, Whistler, British Columbia, Canada* (pp. 765–778). Washington: International Network for Environmental Compliance and Enforcement.

Macken, K. (2011). Strengthening credibility in the EU ETS following security and fraud related incidents. In *INECE 9th International Conference on Environmental Compliance and Enforcement, Conference Proceedings, 20–24 June 2011, Whistler, British Columbia, Canada* (pp. 284–289). Washington: International Network for Environmental Compliance and Enforcement.

Moore, D. (2012). State and local environmental enforcement. In M. Clifford & T. D. Edwards (Eds.), *Environmental crime* (2nd ed., pp. 183–214). Louisville: Jones & Bartlett Learning.

National Association of Attorneys General & United States Department of Justice—environment and natural resources division. (2003). *Guidelines for joint state/federal civil environmental enforcement litigation*. Retrieved 2011, August 22 from http://www.justice.gov/enrd/ENRD_Assets/Guidelines-for-joint-state-federal-civil-environmental-enforcement-litigation.pdf.

O'Flynn, J. (2008). Elusive appeal or aspirational ideal? The rhetoric and reality of the 'collaborative turn' in public policy. In J. O'Flynn & J. Wanna (Eds.), *Collaborative governance: A new era of public policy in Australia* (pp. 181–195). Canberra: The Australian National University, ANU E-Press, Australia and New Zealand School of Government.

O'Leary, G., & Lynott, D. (2011). The effective use of environmental enforcement network in Ireland. In *INECE 9th International Conference on Environmental Compliance and Enforcement, Conference Proceedings, 20–24 June 2011, Whistler, British Columbia, Canada* (pp. 756–764). Washington: International Network for Environmental Compliance and Enforcement.

Organisation for Economic Co-operation and Development (OECD). (2005). *Funding environmental compliance assurance: Lessons learned from international experience*. Paris: OECD.

Pink, G. (2008). Building regulatory capacity in environmental agencies: Through tailored training. In *INECE 8th International Conference on Environmental Compliance and Enforcement: 5–11 April 2008, Cape Town, South Africa, Proceedings* (pp. 225–233). London: Cameron May.

Pink, G. (2010a). *Government co-ordination to enforce environmental laws: The experiences of an Australian environmental regulator*. Canberra: Department of the Environment, Water, Heritage and the Arts, Compliance and Enforcement Branch.

Pink, G. (2010b). *Environmental enforcement networks: A qualitative analysis*. (December 7, 2010). Available at Social Science Research Network. Retrieved 2011, April 14 from http://ssrn.com/abstract=1803179.

Pink, G. (2011). Assessing the utility of environmental enforcement networks: Maximising benefits to members. In *INECE 9th International Conference on Environmental Compliance and Enforcement, Conference Proceedings, 20–24 June 2011, Whistler, British Columbia, Canada* (pp. 797–804). Washington: International Network for Environmental Compliance and Enforcement.

Pink, G., & Lehane, J. (2011). Environmental enforcement networks: Development of a network evaluation matrix. In *INECE 9th International Conference on Environmental Compliance and Enforcement, Conference Proceedings, 20–24 June 2011, Whistler, British Columbia, Canada* (pp. 805–821). Washington: International Network for Environmental Compliance and Enforcement.

Reeve, R. (2002). *Policing international trade in endangered species: The CITES treaty and compliance*. London: Earthscan.

Rhodes, R. A. W. (2006). The sour laws of network governance. In J. Fleming & J. D. Wood (Eds.), *Fighting crime together: The challenges of policing and security networks* (pp. 15–34). Sydney: UNSW Press.

Rittel, H. W., & Webber, M. M. (1973). Dilemmas in a general theory of planning. *Policy Sciences, 4*, 413–439.

Royal Institute of International Affairs. (2008). *The growth and control of international environmental crime: Chatham House workshop, 10–11 December 2007*. London: Royal Institute of International Affairs.

Scanlon, J. (2010). Speech delivered to be 79th Interpol General Assembly, 8 November, Doha, Qatar. Retrieved 2011, June 22 from http://www.cites.org/eng/news/SG/2010/SG_statement_interpol_doha.shtml.

Slaughter, A.-M. (2004). *A new world order*. New Jersey: Princeton University Press.

Socioeconomic Data Application Center (SEDAC). (n.d). *Environmental Treaties and Resources Indicators (ENTRI) Website*. Retrieved 2011, August 22 from http://sedac.ciesin.columbia.edu/entri/.

Sparrow, M. K. (2008). *The character of harms: Operational challenges in control*. New York: Cambridge University Press.

Speth, J. G., & Haas, P. M. (2006). *Global environmental governance*. Washington: Island Press.

Steiner, A. (2007). Foreword. In United Nations Environment Programme. *Networking Counts: Combating illegal trade in ozone depleting substances*. Paris: UNEP.

Suggett, D., & Goodsir, B. (2002). *Triple bottom line measurement and reporting in Australia: Making it Tangible*. Melbourne, Victoria: The Allen Consulting Group. Retrieved 2011, August 22 from http://www.environment.gov.au/sustainability/industry/publications/triple-bottom/index.html.

Taschereau, S., & Bolger, J. (2006). *Networks and capacity*. Maastricht: European Centre for Development Policy Management.

The Harvard Project on Climate Agreements. (2010). *Institutions for international climate governance. Policy Brief 2010-1.* Cambridge, Massachusetts: Harvard Project on Climate Agreements.

Tiernan, A., & Althaus, C. (2005). *Managing out: The public sector in the community. PSM Program 2006.* Canberra: Australian Public Service Commission.

Tomkins, K. (2009). Police, law enforcement and the environment. In *Environmental Crime: a reader* (pp. 515–527). Devon: Willan Publishing.

United Nations Environment Programme (UNEP). (2007). *Networking counts: Combating illegal trade in ozone depleting substances* Paris: UNEP.

United Nations Office on Drugs and Crime (UNODC). (2008a). *Environmental crime: Guidance note to UNODC field offices—18 December 2008.* Thailand: UNODC.

United Nations Office on Drugs and Crime (UNODC). (2008b). *Building capacity to counter trafficking in illegal timber and timber products in East Asia and the pacific region.* [Draft concept]. Thailand: UNODC.

Van der Schraaf, A. (2008). Capacity building in the Dutch inspectorate: Bridging the gap. In *INECE 8th International Conference on Environmental Compliance and Enforcement: 5–11 April 2008 Cape Town, South Africa. Proceedings* (pp. 243–251). London: Cameron May.

White, R. (2008). *Crimes against nature: Environmental criminology and ecological justice.* Devon: Willan Publishing.

White, R. (2010). Globalisation and environmental harm. In R. White (Ed.), *Global environmental harm: Criminological perspectives* (pp. 3–19). Devon: Willan Publishing.

White, R. (2011a). *Transnational environmental crime: Toward an co-global criminology.* Abingdon, Oxon: Routledge.

White, R. (2011b). Environmental law enforcement: The importance of global networks and collaborative practices. *Australasian Policing: A Journal of Professional Practice and Research, 3*(1), 16–22.

Wijbenga, B., Ruessink, H., De Wit, A., & Kapitein, P. (2008). Environmental policing in the 21st century – Background, organisation and experience in the Netherlands. In *INECE 8th International Conference on Environmental Compliance and Enforcement, 5–11 April 2008. Cape Town, South Africa. Proceedings* (pp. 323–333). London: Cameron May.

Williams, P. (2006). The governance of sustainable development in Wales. *Local Environment, 11*(3), 253–267.

Winkworth, G. (2006). Ordinary officials – Building community capacity through cross-sectoral collaboration. *Public Administration Today, 9* (October–December), 31–38.

Chapter 8
Oil Production, Climate Change and Species Decline: The Case of Norway

Ragnhild Sollund

Introduction

In this chapter, I discuss causes for and consequences of climate change, concentrating basically on species decline. With a temperature increase of 2–3°C, between 20 and 30% of the earth's species risk going extinct. In the northern hemisphere species at risk include the polar fox, which I return to, the polar bear, various seal species, a great number of fish stocks and sea birds, such as *Lomvi.* When locally situating consequences of climate change, I find it justified also to locally situate *causes* for climate change and their relation to carbon emissions. This is done in the specific Norwegian context and in light of the ideology underpinning the developed oil industry in Norway, as it is motivated by short- and long-term financial gains. The chapter therefore starts by briefly outlining the Norwegian part of the global oil industry, before turning to some selected harmful effects of this industry related to global warming.

From the point of view of speciesism (see Sollund 2012), the chapter further discusses how a threatened species—in this case, the polar fox—is "saved" from possible extinction in ways that include sacrificing the well-being of individual foxes.

Norway as an Oil-Producing Nation

Norway is an oil and gas producing country. The oil industry in Norway started when it was found exploitable oil resources in the North Sea in 1969. In a ranking of the states in the world with most oil resources, Norway is listed as number 17. According to the Norwegian Ministry of Oil and Energy, Norway is number five of

R. Sollund (✉)
Department of Criminology and Sociology of Law, University of Oslo,
P.O. Box 6706, Oslo 0130, Norway
e-mail: ragnhild.sollund@jus.uio.no

R. White (ed.), Climate Change from a Criminological Perspective,
DOI 10.1007/978-1-4614-3640-9_8,

the most oil exporting countries in the world, with exports amounting to nearly 2.5 million barrels per day. The oil industry has been of huge importance in the development of the Norwegian welfare state and the total amount of the Norwegian oil resources is estimated to NOK[1] 8,000 billion. Roughly 3,000 billion NOK is so far invested in the State pension fund, half a million NOK per inhabitant. The Ministry of Oil and Energy states at its web page[2]: "The total recoverable petroleum resources on the Norwegian continental shelf (NCS) are estimated at some 13 billion standard cubic metres of oil equivalents (*scm o.e.*). There is a high degree of uncertainty in estimating this, and so the total resources are calculated to be somewhere in the interval between 10.6 and 16.9 billion scm o.e. Of these resources, 35% are already sold and delivered, and the remaining 65% are distributed as follows: 28% are proven resources, 11% are contingent resources yet to be decided for development, and 26% are undiscovered resources".

There has been a long public debate about the prospects and consequences of the oil industry, not the least in terms of environmental harms for the fishing industry from which Norwegian vessels delivered 2.7 million tonnes of fish, crustaceans and molluscs in 2010, up 6% compared with 2009. The landed value amounted to NOK 13.2 billion, up 17% from the previous year.[3] Due to the cultural and economical importance of the fishing industry in Norway, in combination with the worries provoked by the BP disaster in the gulf of Mexico in April 2010, The Norwegian government agreed to postpone plans about further oil drilling in the vulnerable parts of Lofoten and Vesterålen in Northern Norway in the spring of 2011, because of the important populations of fish, not the least cod, in this area.

A significant part of the CO_2-emissions in Norway come from the Norwegian continental shelf. According to the Norwegian oil directorate, in 2010, emissions from petroleum activities amounted to 12.6 million tonnes CO_2. This is a small increase from 12.4 million tonnes the previous year. Greenhouse gas emissions from oil and gas activities have been relatively stable over the past 10 years.[4] Most of the emissions come from gas burning in turbines. The drilling and oil and gas production also entails emission of polluted water and chemicals into the sea, endangering marine life with serious long-term effects.[5]

[1] Norwegian kroner. Hundred NOK is €12.89.

[2] http://www.regjeringen.no/nb/dep/oed/tema/olje_og_gass/norges-olje-og-gassressurser-.html?id=443528. Accessed 17 July 2011.

[3] Parallel with the development of the oil industry, those living from fishing have declined in numbers from 68,000 in the 1950s to 10,000 today. (Statistics Norway 2011: http://www.ssb.no/english/subjects/10/05/fiskeri_havbruk_en/. Parallel to this development there has been a huge increase in fish farming, predominantly salmon, which now constitutes 90% of Norwegian fish export. The detrimental effects fish farming has on the environment, on the wild salmon stocks and also in terms on individual abuse and suffering should be subject to attention at another occasion.

[4] http://www.npd.no/en/news/news/2011/oil-and-gas-industry-emissions-and-discharges-2010-/. Accsessed on 18 July 2011.

[5] http://no.wikipedia.org/wiki/Petroleumsvirksomhet_i_Norge. Accessed 18 July.

The Norwegian oil company Statoil, of which the Norwegian state has 67% of the shares, is also involved in the highly polluting oil production associated with the Alberta tar sands in Canada. According to Greenpeace, this oil production threatens water resources and animals, creates conflicts with indigenous groups and releases 13 times more carbon emissions than ordinary oil production. Greenpeace and the World Wildlife Foundation thus suggested in the annual shareholder meeting in May 2011, that Statoil withdraw from the tar sand oil production in Canada, but the state voted against, and the production continues.

Buying Free from Guilt

With the extensive amount of carbon emissions from the oil industry, Norway as a nation greatly contributes to global climate change. Maybe because of this, the Norwegian government is supporting measures to reduce carbon emission caused by deforestation, which again is often caused by illegal felling and trade in tropical timber, bad government in forestation and corruption, as well as civil wars (See Boekhout van Solinge 2008a, b). Norway thus partakes through financial support to the Multi Stakeholder Forestry Programme which was initiated by the British Ministry for International Development. Norway supports projects focusing on the financial structures facilitating illegal deforestation in Indonesia and Brazil, forest management and legislative measurements and law enforcement to prevent illegal deforestation.[6] Norway has agreed to contribute with three billion NOK ($US550 million) per year to reduce emissions from deforestation and forest degradation in developing countries, like Indonesia, where the Norwegian government for example contributes with 35 million NOK in a project against deforestation in Papua New Guinea. Three countries are main recipients of the Norwegian aid against deforestation: Indonesia, Brazil and Guyana. Between 2008 and 2010, NOK 55 billion of the state budget so far has been dedicated to prevent deforestation, however only 293 millions were actually paid out and the remaining approximately 90% of the money is still in bank accounts in Oslo and Washington (Mjaaland et al. 2011). Indonesia has been promised six billion, of which only 185 million so far have been paid out through UNDP. The reason for this inertia is the political situation in receiving states and that the Norwegian government wants to secure that the money is used according to its intention.

Although such contributions may appear invaluable in preventing deforestation, which in addition to carbon emissions and entailing climate change also produces habitat loss and species decline, Norway may be critiqued for polluting with one hand, and making remedies with the other. As observed by Nigel South (2008, p. 191):

> [In response], one fashionable and guilt-saving strategy promoted by western nations had been the idea of buying tropical rainforest to preserve it and reduce destructive develop-

---

[6]http://www.regjeringen.no/en/dep/md/tema/internasjonalt_miljosamarbeid/miljosamarbeid-med-utviklingsland/miljovernsamarbeid-med-indonesia/klima-og-avskoging.html?id=464171. Accessed at 18 July 2011.

> ment. However, backed by Survival International, representatives of the Yanomami tribes have argued this trend "is linked to health and social crisis among indigenous people, including sickness, depression, suicide, obesity and drug addiction"

The Norwegian Prime Minister Jens Stoltenberg's engagement for carbon emission quotas can be criticised for being a rich country's way of legitimating its own pollution. In 1991, Stoltenberg and Norway took the initiative to trade in climate quotas, which is an important part of the Kyoto agreement, through CDM: Clean Development Mechanism, whereby developed states which must reduce their own carbon emissions can avoid this by financing the development of pure energy production and other climate projects in developing states.[7]

The Norwegian government has dedicated NOK four billion to buying quotas. The Norwegian goal thereby is not to reduce the country's own carbon emissions, but to fulfil the obligations in the Kyoto agreement through purchasing quotas on the international market. By buying quotas the Norwegian state can make rationalisations for its own carbon emissions, and the ways in which this is part of a capitalist ideology which is not sustainable in environmental terms. Further, by the focus on "economical development" one increases consumerism and exploitation, i.e. related to the meat industry and textile industry in developing countries from where Norway imports clothing.

Norway is greedily extracting oil resources with one hand and trying to buy the Norwegian state free from guilt by the economical means financed through the oil and gas production, whether in Norway or in other state where Norway is involved through Statoil, e.g. Brazil. Maybe, the Norwegian state could rather apply what Rob White (2011) suggests; namely horizon scanning. What will be the *consequences* of extracting all resources now in such a hurry only to keep the money in bank deposits and shares world wide? Would it not be a better solution to leave the resources where they are, for future generations, to prevent the harm entailed by the oil production at present and in the future? White (2011, p. 32) thus underlines the importance of looking *beyond* the near future to see those issues and trends most likely to involve environmental crime. He says that horizon scanning as an intellectual exercise and planning tool can provide insight into threats and actual and potential problems which at present are poorly recognised, and thus to find ways to mitigate problems.

Norway's Indirect Contributions to Environmental Harm and Carbon Emissions

This question must also be seen in perspective of how the money gained by the oil production/extraction is invested. In addition to securing development and the Norwegian welfare state, as mentioned roughly NOK three billion is invested in the Norwegian Government Pension Fund—Global. The State Pension Fund has been criticised for investing in a large number of unethical companies, in addition to

[7] http://avis.dn.no/artikler/avis/article7361.ece. Accessed 18 July 2011.

lending money to states where human rights are not respected. According to Attack Norway, the reason for the investment in controversial companies is simple. Profit has always been the explicit goal of the State Pension Fund, independent of the consequences for humans and the environment. The establishment of an Ethical advisory board in 2004 has, still according to Attack, not changed this (Gausdal 2010)[8]. Bellona, a leading environmental organisation in Norway, has also criticised the State Pension Fund for preserving and prolonging a situation causing global warming through its investments. They say:

> Currently, the Fund helps to maintain "business as usual" by investing in resource-intensive companies and businesses. Rather than acting as an agent for change, the Fund's investments in companies that are materially contributing to global warming and natural resource depletion help preserve the status quo and escalate our challenges. By maintaining such investments, the Fund becomes a de facto supporter of global warming and the damage it causes. (Hauge et al. n.d.)[9]

The Norwegian State Pension Fund has for example been criticised for buying shares in companies which are responsible for illegal deforestation. According to the Norwegian section of the Rain Forest Foundation, the Norwegian State Pension fund has made investments in oil companies such as Repsol, Occidental and Chevron, the foresting companies, Samling and Olam and the palm oil company Wilmar International. More than NOK two billion have been invested in Repsol, which has been heavily criticised for abuse against extremely culturally and physically vulnerable indigenous groups in Peru and Ecuador.[10] Chevron was in February this year in an Ecuadorian court convicted to pay USD 8.6 billion in damages, 860 million of which is to be paid directly to the Amazon defence coalition, the group formed to represent the plaintiffs.[11]

In addition to the long-term effects of global warming caused by the Norwegian part of oil and gas production, oil production entails dispute and conflicts over land rights in Canada, and indirectly through the oil derived financial investments, e.g. in Chevron and Repsol; Norway is also responsible for the displacement of indigenous people. These crimes/harms are definitely a breach with environmental and human rights, and also combine harms/crimes of pollution with other harms against people, and the environment (South 2008). To further follow White (2008, 2011), this is incompatible with ecological justice, "in which; ecological citizenship acknowledges that human beings are merely one component of complex ecosystems that should be preserved for their own sake via the notion of the rights of the environment" (White 2012). In addition to these harms caused by oil production where the Norwegian State Pension Funds has part of its investments, there is the destruction of habitat for a great

---

[8] http://arkiv.attac.no/nyheter/omskogogtraer/. Accessed on 26 July 2011.

[9] http://bellona.org/filearchive/fil_bellona_statement.pdf. Accessed on 26 July 2011.

[10] http://www.regnskog.no/hvordan-vi-jobber/forbrukersp%C3%B8rsm%C3%A5l/trekk-ut-oljefondet. Accessed 22 July 2011.

[11] http://www.energydigital.com/sectors/chevron-texaco-lawsuit-ecuador-court-rules-environmental-damages. Accessed 20 July 2011.

number of species which live in the rainforests, thus causing species going extinct, loss in biological diversity and ecological degradation. The focus in the following section is, however, on the indirect consequences of oil production in terms of climate change, which in turn causes the destruction of habitat for a number of species, in the Norwegian, local context.

Local Consequences of Climate Change: Species Decline

At a local, national level, climate change endangers several species in Norway, as seal and whale species and the polar fox, which is of specific interest because of the counter measures set in to prevent the extinction of this species. Species become endangered e.g. as a consequence of lack of suitable living areas and competition with spreading species, as when the polar fox is displaced by the red fox. In addition, living conditions for the polar fox are affected by the access, or lack of access to prey, as lemmings, which is climate dependent. Lemmings normally came in great numbers with regular intervals; however with the exception of this year—2011—this has not happened since the 1990s. The reason is that mild winters deprive them of a place to live and breed as they live under the snow where they feed on moss.[12] When the snow gets too heavy they cannot produce the tunnels they depend on in the snow. The polar fox in Norway is critically endangered, from 1998 to 2008 there were in total 241 breeding litters in Norway *and* Sweden, of which 111 were in Norway. In 2009, no breeding litters were documented in Norway, as there was a collapse in the population of small rodents and lemmings in Norway in 2008 and 2009. The great decline in the number of polar foxes in Norway led to measures being made in an attempt to save the species by The Norwegian Directorate for Nature Management, through a breeding programme ran by the Norwegian Institute for Nature Research (NINA).

The Polar Fox Surveillance and Breeding Programme

The project was started in 2003 and includes counting and surveillance, including genetic surveillance, of the nests of the polar foxes in Norway, and also a breeding programme through which polar foxes are bred in captivity and the cubs are placed in nature (Eide et al. 2008). The breeding programme for polar foxes was established in 2005 to re-establish, strengthen and tie together Scandinavian populations of polar foxes and to increase genetic exchange and counteract genetic isolation.

[12] http://www.newscientist.com/article/dn19982-plagues-of-lemmings-driven-by-winter-breeding.html. Accessed 18 July 2011. http://www.forskning.no/artikler/2011/januar/276398. Accessed 18 July 2011.

There is a breeding station from which cubs are introduced and the programme relies on capturing cubs in nature from the Scandinavian groups to breed on them. So far in 2011, nine litters with polar foxes have been born, all together more than 85 cubs on Dovrefjell.[13] NINA has since 2007 set out 76 cubs in Dovrefjell which have been bred at the breeding station in Oppdal. In 2010 the first litters of free born cubs were born since the project started, and altogether 39 cubs divided on five litters have been born. The project manager at NINA attributes the success to a combination of the food stations for the polar foxes which have been established, and the good lemmings' year.

Despite the success in reintroducing the polar fox species to Dovrefjell where it had been extinct for nearly a hundred years, the programme has also been subject to critique, due e.g. to the mortality rate of the released foxes. The research team says in their annual report from 2010 (Landa et al. 2010) that of the 87 animals which have been released during the years 2006–2009, only a total of 56 of these animals were found in the collected data within the first year after release, and 32 of these remained in the data collected during 2010. This may indicate that more than half of the released individuals have died. Only 26 of the 38 animals released in 2009 were found in the data collected in 2010. Two of the "recaptured" animals are definitely confirmed dead. According to the researchers the foxes are hard to track as they wander, and when and how they die can be hard to establish, despite them being collared.

Species Survival Versus Animal Abuse: Speciesism

As the foxes which are released depend on humans to feed them and are bred in captivity, an interesting dimension worthy of discussion appears: Is the programme really reintroducing a wild species into Norwegian nature, or is it, despite of the efforts in collecting genetically varied breeding pairs, rather introducing semi-domesticated animals which depend on humans for survival into Norwegian wilderness where the mortality risk is high?

Critique has thus been raised against the programme by animal welfare organisations[14] for putting animals out to suffer and die, thereby seeking to ensure species survival rather than protecting individuals. The Animal welfare alliance finds that the entire project is based on animal abuse, claiming that 50–75% of the animals will die, as they have not been socialised into survival in the mountains. The researchers' response to the critique is that despite the survival of only 25 individuals in 2008–2009, the high mortality was caused by the lack of small rodents which also caused all free

[13] http://www.nina.no/Aktuelt/Artikkel/tabid/945/ArticleId/1451/Historisk-mange-kull-med-fjellrev.aspx. Accessed 18 July 2011.

[14] http://www.njff.no/portal/page/portal/njff/nyhet?element_id=101099931&displaypage=TRUE. Accessed 19 July 2011.

born litters to die. However; this may again question the viability of the programme; as the main reason why rodents fail to breed, at least in the case of the lemmings, is mild winters and lack of snow, and despite of the claimed success of the breeding programme, the foxes will continue to depend on humans for survival.

The project has further been criticised for the killing of wild polar foxes which lived at Finse. The reason why the animals were killed was that this group had mixed with escaped farm foxes, and consequently, was genetically "impure". Despite protests, for example by the Council for Animal Ethics, the animals were killed.[15] This decision made by the Ministry of Environmental affairs seems paradoxical when the project also puts much effort in capturing wild animals from different populations in order to achieve genetic variation. In this case at Finse, the foxes themselves had managed both to secure genetic variation as well as species survival, which is the explicit goal of the breeding programme. In order to achieve this, however, a number of foxes from the critically endangered species have been caught through painful, abusive methods, and have later died in captivity because of stress. The "genetic impurity" of the Finse group was actually discovered when they were caught to provide breeding material for this part of the project.

From an individual perspective, for a cub to be released into the mountains where s/he will suffer from starvation, will entail suffering, and as shown often, death. To follow Piers Beirne (1999, 2009) in his discussion of animal abuse, such acts should be acknowledged as abusive and painful, and thus meriting the same attention which is directed to abuse when humans are the victims. Such abusive acts against non-humans cannot be disconnected from speciesism—the practice and ideology of systematically discriminating other species, most often for some kind of human benefit (e.g. Singer 1995; Regan 1983; Nibert 2002; Noske 1989). Speciesism should not exclusively be understood as discrimination against non-human species, but more importantly against the *individuals* of non-human species. The acts directed to many of the individual foxes through the Norwegian breeding programme, can be characterised as speciesist abuse, as humans through the project physically and mentally harmed the animals through captivation of former free individuals and also through the release of those being captives, for human defined purposes. As more than 50% of the released cubs died, as well as several of the initially caught foxes, they were also victims of theriocide (the animal equivalent of homicide) (Beirne 2007, 2008). One aspect of this is for example that the individuals which were released were labelled with collars in order to identify them, with the harmful effects this has for the individual which must wear it. The researchers take blood samples of the foxes, with the anxiety this must entail for a fox who is not accustomed to humans. Research on wolves which are subject to the same kind of procedure shows in this respect that wolves who have once been subject to human examination and labelling, for ever will try to avoid their human molesters (Tønnessen 2010). This form of control over "wild" animals is abusive and part of speciesist practice, something

[15] The Council for Animal Ethics is an independent advisory body appointed by the Ministry of Agriculture and Food in collaboration with the Ministry of Fisheries and Coastal Affairs.

which becomes evident if one imagines putting humans in a similar situation. Cazaux (2007) finds for example that the collars often directly and indirectly can influence on the collared animals' chances of survival, and not the least, well-being. She says: "As the mere presence of humans—however low profile this might be—potentially influences the animal's life in a harmful way, handling them and marking them can derivatively be presumed to have a negative impact on their lives" (Cazaux 2007, p. 101).

The Directorate for nature management in Norway is not a green movement, but part of the state apparatus. One of the aims of the Directorate is, however, to secure biological diversity and thus that Norway acts in accordance with the Convention of biological diversity and the Berne convention, both signed by Norway. As for green movements, the logic is to secure *species* rather than individuals, and Norwegian authorities can thus also be criticised for being anthropocentric, in ignoring individual well-being and inflicting animals with harm and in advancing human interests for preserving the polar fox as a "wild species".

Svärd (2008) has shown how green movements often can be accused of speciesism, as for green movements other animals have value as *species*, not as individuals. He states that environmentalists and conservationist NGOs as representing the extra-parliamentary leg of the Green movement have focused their attention on other animals mainly on a species basis, and not based on individual animal rights, as these are advocated for example by Tom Regan (1983). The species category disguises that animals are individuals with individual rights and the rights view does not recognise the moral rights of species to anything, including survival (Regan 2004, p. 359, here in Svärd 2008, p. 172). In such analysis, the species category is but an analytical, zoological category and as such, cannot be hurt or harmed. As a species goes extinct however, this is often the consequence of individual suffering on an accumulated level, for example when individuals die of starvation, for example when they cannot find food due to species decline caused by climate change.

Species Justice and Individual Animal Justice

This implies that a species cannot only be reduced to an analytical category, and animals cannot be reduced to being only part of a species, but must be perceived as individuals. In contrast to green movements, and the logic to be found for example in the CITES convention, through which it is clear that non-human species have not individual rights, but only rights as part of a species, Svärd underlines: "From an animal rights perspective, nonhumans are entitled to concern and respect as individuals, and this entitlement may never be dependent on the remaining size of the rights-holder's [a species] group" (Svärd 2008, p. 172). However, practices which threaten and harm individual rights, as a consequence may also threaten the survival of species, though it is not the species' rights which should be prioritised, but the individuals which together form a species. Species rights, should thus be seen as a prolongation of individual rights, rather than just seeing individuals as categorical representatives of a species, as in the CITES convention, when referring to individuals as "specimens" (Sollund 2012). Species justice may be the outcome of individual justice, the opposite is not necessarily guaranteed. When individual animals' species dependent needs are met, and individual harm is not inflicted, than species may survive, and species justice can be accomplished.

One question is whether species justice (not individual justice) can be accomplished when the species can survive only by means of "artificial" feeding from humans, and through human orchestrated breeding programmes and under human control. Can the species at all be claimed to survive within the frames of this species natural characteristic feature, abilities and needs, when individuals become dependent of and accustomed to, human assistance, somehow turning them into patients? Maybe, as a consequence, one could claim that the species then cease being this particular species, as only the genetic "material" remains, and not the animals' historically constituted practices, such as those related to food hunting. And again, as the species is compounded by individuals who suffer from this kind of deprivation of liberty and painful measures, turned them into someone different from those they should have been. Consequently, may such measures made for species preservation thus be in vain and counteracting their intention? A question to follow is also, is it at all fruitful to talk about species justice as a phenomenon as long as a species (in some form) can survive, despite or maybe even only by means of the infliction of harm directed to this species' individuals? In short: Can justice be done to a species, when harm is inflicted on the individuals forming the species on a systematic level?

This can be seen in White's perspective as he categorises individual animal rights and species justice as interdependent, when listing these two phenomena as one of the broad approaches to justice identified in green criminology: "in which environmental harm is constructed in relation to the place of non-human animals within environments and with their intrinsic rights to not suffer from abuse, whether this be one-on-one harm, institutionalised harm, or harm arising from human actions that affect climates and environments on a global scale" (White 2011, p. 23). This also echoes Beirne and South's perspective of green criminology stating that it is difficult to disentangle environmental harms from the abuse of non-human animals. "Animals of course live in environments and their well-being—physical, emotional, psychological—is absolutely an intimately linked to the health and good standing of their environments" (Beirne and South 2007, pp. xiii–xiv).

In this perspective, the killing of polar foxes on Finse because they had wrong genes can be defined as a breach of individual rights, and consequently a breach of species justice as well as environmental harm, as polar foxes are part of the environment. In my opinion, a distinction must however be made between those harms which are directly inflicted upon individuals, and those which are the indirect consequences of e.g. pollution, climate change, deforestation and loss of habitat.

This became particularly salient as the theriocide of the Finse group followed the logic of a eugenics programme when directed to humans during the Nazi period. In regarding the Finse group of polar foxes as contaminated by the farm foxes, it becomes evident that the farm foxes are inferior to the "wild" polar fox, and therefore must be exterminated, except in the farm cages where they are kept to "produce fur".

The alienation by the humans necessary to commit the atrocities of encaging wild animals is interesting in a speciesist perspective per se (see Sollund 2008). However, this case is also interesting because it appears that is was the "wildness"—the absence of genetic "contamination" of the "genuine" polar fox—and thus also the absence of human influence and contamination through breeding programmes and incarceration of the "farm polar fox", which give the "wild" foxes their value, and not, of course, the inherent value of sentient beings with proper interests and rights. This is also interesting as it is through fox eugenic programmes that the "fur fox" is cast in his/her victim role, whereas the "wild fox" ironically and paradoxically also through breeding programmes, shall be preserved as a "wild species". Consequently; it is the way the foxes in different forms provide "a good" for humans, either in financial terms, as "fur" or for recreational purposes as a "token of nature and wilderness", that makes it possible to breed them in different directions, though interrelated as they are both subject to human control.

As the Nazis ranked human "races", and white, male humans have categorised races and sexes according to their value based on different kinds of measurements, e.g. of their brains and the placement of their navel (Gould 1981) the Norwegian fox researchers determined that because of the "contamination of their genes" the foxes should die. In this, they performed a "double speciesist act" of theriocide. Not only could they, as humans, determine that the foxes should die, but they would also rank the different fox breeds according to their "wildness genes" and thus their genuineness and consequently kill those regarded as worthless, precisely because of the human regime they had been subject to.

Conclusion

In this chapter, I show the harmful effects of legal activities which raise questions for further discussion, in terms of how such harms should me met. I show the interrelatedness between the exploitation of oil resources in Norway and their direct and indirect harmful environmental consequences. I further assess the seeming incompatibility of making further profit from the oil industry by making investments through the Norwegian State Pension Fund in international, capitalist enterprises, e.g. in the oil sector on one side, and the preservation of the environment, the respect for human as well as animal rights on a global level on the other. As the production of oil and gas entails carbon emissions to a high degree, the indirect long-term consequences can already be visible in terms of climate change in Norway, which in turn affects the survival of a number of species, exemplified by the polar fox. An anthropocentric stand to this is seen through the attempts to secure species survival of the polar fox, whereby disrespecting individual animal rights through the theriocide of unwanted animals with "impure genes", and breeding programmes leading to semi-domestication of the foxes. This example shows an approach that takes into account only the survival of the species and shows that who merits to be reckoned as part of a species worth of protection is determined by humans and is also under human control. The actions involved in such an approach are anthropocentric and speciesist. This suggests that rather than just regarding the species as a *category* worthy of protection, one must start by the respect for the individuals compounding a species. In order to do justice to a species, one must take into account the environmental, social, physical and psychological needs of individuals with similar needs and features, those comprising a species. The best way to achieve this is quite simple: To stop the destruction of habitat which deprives non-human "wild" species of the natural conditions they need to live and breed, and allow the individuals of each species a life in freedom, free from human restrictions. Regarding domesticated species, they should not be regarded as means to an end, but valued as individuals with personal needs and desires according to their species' needs. Imprisoning animals is not according to their needs and incompatible with species justice. An unrealistic utopia which would be the consequence of genuine species justice would therefore be to abstain from practices which depend on the exploitation of other animal species. In the meantime, a minimum for the animal slaves would be to allow them necessary space, adequate food and access to fresh air and offspring.

This chapter would invite to further examination of environmental harms caused by states' and corporations' legal actions, to assess harms and how these can be counteracted, and to further develop the theoretical framework of green, eco-global criminology. One issue which should be subject to further attention could for example be how to operationalise terms such as environmental justice, ecological justice and species justice (White 2008, 2011). Does it for example make sense to talk about species justice when referring to humans, or do we when regarding the human species only find it relevant to talk about human rights? And if so, could this approach, if applied to other species also lead us to give them justice? One dimension of this could be to further examine the ways in which different aspects of green crimes interrelate, such as the disrespect of human, animal and ecological rights, which can be hard to disentangle from racism, speciesism and anthropocentrism.

References

Beirne, P. (1999). For a nonspeciesist criminology. Animal abuse as an object of study. *Criminology. An Interdisciplinary Journal. The official publication of the American Society of Criminology, 1*, 117–150.

Beirne, P. (2007). Animal rights, animal abuse and green criminology. In Beirne, P., & Nigel, S. (eds.), *Issues in green criminology*. Devon Willan, 5587.

Beirne, P. (2008). *Confronting animal abuse. Law, criminology and human-animal relationships.* Rowman and Littlefield.

Beirne, P. (2009). *Confronting animal abuse*. Lanham, MD: Rowman and Littlefield publishing.

Beirne, P., & South, N. (2007). Introduction to green criminology. In *Issues in green criminology. Confronting harms against environments, humanity and other animals* (pp. xiii–xxii). Devon: Willan Publishing.

Boekhout van Solinge, T. (2008a). Crime, conflicts and ecology in Africa. In Sollund, R. (ed.), *Global harms ecological crime and speciesism* (pp. 13–35). New York: Nova Science Publishers.

Boekhout van Solinge, T. (2008b). The land of the orang-utan and the bird of paradise under threat. In Sollund, R. (ed.), *Global harms ecological crime and speciesism* (pp. 51–71). New York: Nova Science Publishers.

Cazaux, G. (2007). Labelling animals: non-speciesist criminology and techniques to identify other animals. In *Issues in green criminology, confronting harms against environments, humanity and other animals* (pp. 97–114). Devon: Willan Publishing.

Eide, N. E., Øystein, F., Roy og Landa, A., & Arild. (2008). *Fjellrev i Norge 2008. Resultater fra det norske overvåkingsprogrammet for fjellrev. [Polar foxes in Norway 2008. Results from the Norwegian surveillance program for polar foxes.] Norsk institutt for naturforskning. (NINA) Rapport 389.*

Gausdal, L. (2010). Om skog og trær [About forest and trees] *Dagsavisen 20.12.2010.*

Gould, S. J. (1981). *The mismeasure of man.* Baskerville: Penguine.

Hauge, F., Birgitte, L., & Eivind, H. (n.d.) *Bellona's recommendations for the ethical guidelines of the Norwegian Government Pension Fund—Global.* http://bellona.org/filearchive/fil_bellona_statement.pdf.

Landa, A., Mari, T., Roger, M., & Roy, O. A. (2010). *Avlsprogrammet for fjellrev. Årsrapport 2010.* NINA rapport 603. Retrieved 19 July from http://www.nina.no/archive/nina/PppBasePdf/rapport/2011/603.pdf.

Mjaaland, O., Nils-Inge, K., & Ann, F. (2011). Millarder står ubrukt. Uenighet sinker regjeringens pengestøtte til vern av regnskogen. [Billions unused]. Disagreement delays the government's financial support to save rainforest' *Aftenposten 19.07.2011.*

Nibert, D. (2002). *Animal rights Human rights.* Oxford: Rowman and Littlefield publishers.

Noske, B. (1989). *Humans and other animals: Beyond the boundaries of anthropology.* London: Pluto Press.

Regan, T. (1983). *The case for animal rights.* California: University of California Press.

Regan, T. (2004). *The case for animal rights.* University of California Press.

Singer, P. (1995). *Animal liberation.* London: Pimlico.

Sollund, R. (2008). Causes for speciesism. Difference, distance and denial. In Sollund, R. (ed.), *Global harms. Ecological crime and speciesism* (pp. 109–131). New York: Nova Science Publishers.

Sollund, R. (2012). Speciesism as doxic practice. In R. Ellefsen, G. Larsen, & R. Sollund (Eds.), *Eco-global crimes: Contemporary and future challenges.* London: Ashgate.

South, N. (2008). Nature, difference and the rejection of harm: expanding the agenda of green criminology. In Sollund, R. (ed.), *Global harms ecological crime and speciesism* (pp. 187–200). New York: Nova Science Publishers.

Svärd, P.-A. (2008). Protecting the animals. An abolitionist critique of animal welfarism and green ideology. In Sollund, R. (ed.), *Global harms ecological crime and speciesism* (pp. 167–187). New York: Nova Science Publishers.

Tønnessen, M. (2010). *Ulovlig jakt på ulv* [Illegal wolf hunting] Speech at Kriminalpolitisk seminar [Criminal political seminar] University of Oslo, Dept of Criminology and Sociology of Law. 11.11.2010.

White, R. (2008). *Crimes against nature: Environmental criminology and ecological justice.* Devon: Willan Publishing.

White, R. (2011). *Transnational environmental harm.* London: Routledge.

White, R. (2012). The foundations of eco-global criminology. In R. Ellefsen, G. Larsen, & R. Sollund (Eds.), *Eco-global crimes: Contemporary and future challenges.* London: Ashgate.

Chapter 9
Climate Change, Gender and Natural Disasters: Social Differences and Environment-Related Victimisation

Diane Heckenberg and Ingrid Johnston

Introduction

Climate change is likely to bring with it an increase in the frequency and/or severity of natural disasters such as flooding, heat waves and cyclones. The combination with other impacts such as rising sea levels will intensify these events. Disasters exacerbate existing inequalities and vulnerabilities in the community, meaning that there will be a range of effects for men and women, adults, youth and children, in the developing and Western worlds. Opportunities for such crimes as sexual assault and exploitation of women and children increase, and situational factors such as overcrowding and stress leading to domestic violence will also rise. Natural disasters therefore bring many challenges, but few of them are unique, and many of the recommendations for reducing harms to victims align well with the crime prevention principles of criminology.

It is said that existing social inequalities and vulnerabilities will shape the outcomes of events such as disasters (Mutter and Barnard 2010). It follows then that disadvantages and discrimination faced by women and children in varying forms and to varying degrees throughout the world will be exaggerated by the onset of natural disasters. Indeed, the impacts are different for men and for women, for adults and for children, for boys and for girls, and these are largely a reflection of the existing society and culture. Those who do not know how to swim will fare worse in a flood that those who do. So if a cultural norm is for males but not females to learn to swim, the potential for a higher death toll among women as well as children is in place long before the rains

D. Heckenberg (✉) • I. Johnston
School of Sociology and Social Work, University of Tasmania,
Private Bag 17, Hobart, TAS 7001, Australia
e-mail: Diane.Heckenberg@utas.edu.au; Ingrid.Johnston@utas.edu.au

R. White (ed.), Climate Change from a Criminological Perspective,
DOI 10.1007/978-1-4614-3640-9_9, © Springer Science+Business Media New York 2012

begin. This chapter looks at some of the different experiences of males and females, and of adults, youth and children in natural disasters across the world.

References to gender as an issue in climate change debates and international protocols on disaster management in particular have been made since at least the 1994 Yokohama World Conference on Natural Disaster Reduction, with efforts to empower women and include them as well as men in all stages of disaster management programs (Hannan 2009). Despite this, there is a recognised lack of research and evidence about climate change and gender, related to the paucity of research about climate change and social issues generally, making it difficult to identify and understand the complex links (Terry 2009). This may be no coincidence as it has been suggested that women's voices are more likely to be heard when the social consequences of climate change and variability are debated, but these debates are rare (Alston 2011). We do know that climate change is likely to hamper efforts to achieve both the Millennium Development Goals as it contributes to the cycles of poverty; and the Convention on the Rights of the Child which includes rights to life, survival, development and health (Arts 2009). The current state of knowledge regarding risks particular to men and women, male and female youth, girls and boys is outlined in this chapter, along with some unanswered questions and issues for the future.

Social Vulnerability and Extreme Events

Long-term climate change is likely to have ramifications for gender relations, as well as agricultural, ecological and human systems (Nelson et al. 2002, p. 51). Much of the knowledge about the gendered impacts of disaster is contextualised by humanitarian and natural disasters. The gender lens falls disproportionately on women and girls in the developing world and to a lesser extent, marginalised populations in the developed world, such as the African American women and girls of Katrina. The environmental lens falls disproportionately on agriculturally based communities suffering slow-onset prolonged events such as drought, crop failure and water insecurity or communities affected by rapid-onset natural disasters like earthquakes or extreme-weather events such as hurricanes, cyclones and tsunamis. Less attention is paid to co-occurring natural and technological disasters, like the recent Tohoku earthquake and tsunami in Japan (but see Chap. 11). While much can be learned from the plight of women and girls in these circumstances, comparable attention is not given to the perspectives, priorities, needs and experiences of adult men, male youth and boys. This is addressed in the latter sections of this chapter.

Gender relations place men and women differently at risk in the face of climate change, extreme weather events and natural disaster: "Gender shapes men's interactions with men as well as women during crises, and differently in different contexts" (Enarson 2009, p. 1). Gender expectations can be self-imposed, reinforced by intimate partners, family, friends, work colleagues and peers as well as being embedded in the social norms of particular communities. Gender norms are reproduced by the institutions of the state, by stereotypical images in the mass media especially during high-profile events, and by the discourses surrounding climate change, disaster and environmental degradation. Gender norms and values and resulting behaviours can have negative effects (e.g. on health), but gender norms and values are not fixed, can evolve over time, vary substantially from place to place, and are subject to change (WHO 2010, p. 2).

Gender difference and gender bias also occur in the discourses surrounding climate change and environmental degradation. An example of this is use of the prefix "man-made" to describe events (e.g. "man-made" climate change, disaster), environmental harms (e.g. "man-made" greenhouse effect, environmental degradation), and substances (e.g. "man-made" chemicals). Inherent in this language is an implication that men are responsible for all of these scenarios. Yet, women have been passive and active partners in the world's factories, fields, and farms. Alongside men, they have applied pesticides and insecticides to fields, orchards and farms with the residue being released to rivers and streams; disposed of chemically-laden products including plastics, household cleaning products, personal care products and thousands of tonnes of disposable nappies, ultimately contributing to bulging landfills, greenhouse gas emissions and long-term climate change.

As Brody et al. (2008, p. 2) note "a gendered approach to climate change should not simply be about women and girls; men and boys are also vulnerable to the impacts of climate change, but often in different ways. Furthermore, women and girls are involved in relationships with men and boys and it is at the level of these gender relations and the social expectations influencing them that research needs to be conducted and change needs to happen". Greater clarity is needed to better understand specifically how gender roles change in response to different types of events. In this context, it is the notions of *change* and *conflict* that emerge as connecting threads. For example, "Gender norms are challenged when women take on tasks traditionally ascribed to men, gaining new skills and changing prevalent views as to women's capabilities. This occurred after Hurricane Mitch when women were observed building shelters and wells" (PAHO 2001, p. 2 cited in Nelson et al. 2002, p. 56).

In the aftermath of a disaster, everything changes. Disaster forces people who may have been taking such necessities as food, water and shelter for granted, abruptly down Maslow's hierarchy of needs, to a search for these basics of human life (Olson and Gawronski 2010). In many societies throughout the world, these are the very things that are seen as the responsibility of women (with the possible exception of shelter), and the burden of this also often falls to women.

After a natural disaster, with issues such as grief for loved ones who died and loss of housing, possessions and income, stress levels and tensions increase. This in turn increases the risk to both women and children of abuse. With domestic violence being blind to race, ethnicity and socio-economic status, this increased risk occurs worldwide. While statistics on domestic violence are notoriously difficult to obtain at the best of times, the aftermath of a disaster adds another layer of complexity to the issue. But from the evidence available, reports of increases in domestic violence, coming from those who work in shelters and support services, as well as from victims themselves, are found throughout the world (Houghton 2009; Saroor 2009; Action Aid Nepal 2007; Thornton and Voigt 2007). For example, in the USA, following a hurricane, rates of inflicted head injury to children under 2 years of age increased fivefold (Bartlett 2008).

Impacts on men and women will differ during the relief and reconstruction phases. Cultures and societies that only recognise male-headed households leave women and children who find themselves suddenly the head of their household as a result of the disaster in a position of discrimination. They may have restricted or no access to compensation and disaster relief money from Governments, or to aid which is distributed through the head of the household. There is a need for aid agencies to be able to officially recognise children (as well as women) as the head of a household (Delaney 2006). This recognition also needs to extend to male youth-headed households.

Where people are displaced as a result of natural disaster, there can be flow-on effects for communities outside the immediate area. A community in North Queensland, in Australia, where flooding is an annual occurrence, reported tensions arising between the community evacuated and the "host" community that finds itself with a large group of people camped on the local sporting oval for weeks at a time. The host community blamed the evacuees for any negative event, and the evacuees found their usual capacity for resilience impeded by being away from home (Cottrell 2008). These tensions sow the seeds of conflict between the communities. Sometimes the effects of a disaster can manifest themselves in a more subtle way. A study of communities affected by Hurricane Katrina in the USA found that there was an increased level of bullying among children after the disaster (Terranova et al. 2009).

Victimisation and the Social Impact of Disasters

The majority of case studies on the impacts of disasters use the "mega-disasters" as their focus, and thus many come from the 2005 Hurricane Katrina in the USA, the 2004 Indian Ocean tsunami in Sri Lanka, Thailand and Indonesia and the 2005 earthquake in the Kashmir region.

Hurricane Katrina

Natural disasters create a sudden high demand for labour for the reconstruction efforts. Typically this labour is low wage, hard physical labour—precisely the circumstances in which people trafficking is most likely to occur. There have been nine human trafficking cases identified in the Gulf Coast region, involving 1,383 victims from India, Thailand, Peru, Bolivia, Dominican Republic, Brazil, Mexico, Honduras, Philippines and Panama, starting as soon as 2 months after Hurricane Katrina (Hepburn and Simon 2010). Victims have reported being forced to live in hurricane damaged buildings full of mould and debris, with no electricity or running water. Cooking was done with the available contaminated water, and traps were set for pigeons so there was some food to cook (Hepburn and Simon 2010).

The immediate aftermath of Katrina saw police called off search and rescue missions to control looting and fires (for a critique of the police response, see Chap. 10). In the midst of this societal breakdown were brutal gang rapes (Thornton and Voigt 2007). These events occurred in a variety of settings, some even before the hurricane actually hit. A woman trying to hitch-hike to get out of New Orleans in a hurry during the warning phase was picked up by two men who raped and abandoned her. After the event, a woman was looking in a convenience store for medicine for her sick mother, and food and water for her own children. She was gang raped in the store by a group of young men, frustrated at being unable to break into the bank cash machine (ATM) located there. Several reports were made of men sexually assaulting women they were offering "assistance" to. They were not necessarily strangers—a woman living in her car after being left homeless by the hurricane was offered a place to stay by a co-worker who then raped her (Thornton and Voigt 2007).

Kobe Earthquake

There were reports of sexual assaults and domestic violence following the Kobe earthquake in Japan. Broken street lights created dark corners, and collapsed buildings were the venue for some rapes and sexual assaults on women. Shelters were also considered unsafe, but with women reluctant to report because they felt they had nowhere else to go. Domestic violence was also reported, including by women who took their stress out on their children (Masai et al. 2009).

Whakatane Floods

In one example from a lesser known disaster of a smaller scale in the Western World, in July 2004, the town of Whakatane in New Zealand flooded, leaving it cut off for 2 days, and with 300 homes evacuated. The New Zealand police reported an increase in domestic violence call outs from July through until September, and a women's refuge reported a tripling of its workload. At one point, the safe houses were so overcrowded that each worker had three families staying with them as well (Houghton 2009).

Pakistan Earthquake

Following the 2005 earthquake in Pakistan, examples arose of fear or actual abuse in tent camps. A woman living in a tent tied her 4 year old to the bed because she did not want him to play in the mud which would then require her to wash and dry the clothes in the frigid weather. There was a reported general sense of fear about the possible abduction of women and children, which was severe enough for some men to be unwilling to leave their tents to go to work. A woman living in a tent with her four children stitched the tent flap closed each night because she feared for their safety if people realised they did not have a male "protector" (Sayeed 2009).

Indian Ocean Tsunami

Having the responsibility for caring for children during a natural disaster can bring unique challenges and loss. A Sri Lankan mother reported of the 2004 tsunami:

> When the wave came, I grabbed both by children in my arms and tried to run, but the wave caught me, and I was forced to choose between my six-year-old and my baby or else we all would have perished. I can still see the look in his eyes when I let go of him (Chew and Ramdas 2005, cover page)

Interviews with staff members of non-government organisations, UN agencies and government representatives in Sri Lanka 5 months after the tsunami confirmed incidences of violence including domestic violence, and sexual exploitation that continued in camps and temporary shelters for a period of months. There were reports of sexual assaults in poorly lit toilets; men offering impoverished victims money or goods for sex; women with learning disabilities being attacked as they went from the camp to their damaged houses to look for belongings; sexual harassment of women while they were bathing, including by humanitarian workers; and incidents of rape, physical abuse, sexual harassment and exploitation committed by police and security at the camps (Fisher 2010).

Haiti Earthquake

On 12 January 2010, an earthquake struck Haiti, killing over 220,000 people, and displacing over 1.3 million more. In a country where over 40% of the population are children, it was a time of highly increased vulnerability for those children. One 10-year-old girl whose father was killed in the earthquake was living in an Internally Displaced Persons (IDP) camp with her mother and three younger siblings. This girl reported having been raped twice—once on the way to the communal bathroom in the camp, and once in her tent when her mother was out looking for work (Todres 2011).

Social Context and Physical Location

The available evidence shows that impacts differ depending on the type of event, and where it occurs. In countries that are highly gendered, women and men have distinct roles and responsibilities in daily life. This to a certain extent dictates where they are likely to be when an event occurs, how much warning they may receive, and what their available options are for survival both during and afterwards. Before an event, where there is some warning, it must be communicated throughout the area likely to be affected. Access to these messages will be different, especially in some cultures, for men and for women, for adults, for teenagers, for children, and for the elderly. Inequalities may exist from the outset in who has the knowledge about a coming risk.

Having the information though is not the whole story—people then have to be able to make decisions based on that information, and act on those decisions. Worldwide, young children will need assistance to do both, and elderly adults may not be physically able to act on their decisions without help. In some societies, women also will require assistance with either or both, where cultural norms dictate for example that decisions are made by the males in the household, or that women must be accompanied in public by men. In some areas of Bangladesh, the observance of the socio-cultural norm of purdah (the practice of preventing men from seeing women) is considered more important than disaster risk (Ariyabandu 2009). A cyclone in 1991 in Bangladesh claimed the lives of many women forced by cultural norms to wait at home for their husbands or a male family member to direct them to safety (Saito 2009). This restricted mobility and autonomous decision making puts not only the women but their dependent children at increased risk, a fact recognised in the Kashmir earthquakes in 2005 (Mehta 2009).

Sometimes the differential victimisation of men and women is less obvious to those outside the situation. A group of women in the North West Frontier Province of Pakistan reported being blamed for the 2005 earthquake that was seen by some in the region as not being a natural disaster, but brought on by the "sins of women".

> People blame us for the tragedy. If our moral practices had been better this would not have happened…It is terrible enough to have survived the earthquake, to face the huge loss of family and friends, and possessions; on top of that to feel guilty for all that is being said about our "sins" drives us to despair (Sayeed 2009, p. 145)

While violence and abuse of women and children may be a worldwide phenomenon, after a natural disaster, as at anytime, there are different manifestations of this according to where the event has occurred. Most reports from Western countries identify incidents of domestic violence and sexual assault of women. In developing countries though, more reports also include the sexual exploitation of both women and children. This exploitation and violence takes different forms for boys than for girls. Girls are likely to experience gender based violence aligned with their comparatively powerless position in society, whereas for boys, especially in conflict situations, the victimisation is more likely to be about intimidation (Delaney 2006).

Children are especially vulnerable when they are without an adult. This may be because they have become separated from their parents or caregivers during the event, because they have become orphaned by the event, or because the lone adult in their household is obliged to leave them alone in order to earn an income. Experience from Sri Lanka, Indonesia and Thailand following the 2004 tsunami found that children were vulnerable to sexual abuse and exploitation at all stages of the disaster. In the immediate aftermath there were abductions, sexual assaults and trafficking of children. During the relief and recovery stage the desperation for survival left children vulnerable to prostitution, and in the reconstruction phase, child sex tourism became a significant risk as people searched for long term means of generating income (Delaney 2006).

The situation is made worse when there are no processes in place for responding to reports of abuse and violence. An earthquake generated tsunami in April 2007 in the Pacific nation of the Solomon Islands left about 4,000 families homeless. Between April and October 2007, there were at least 14 officially reported cases of gender-based violence, including rape. However, there were no effective systems in place for referral and response to these reports (OHCHR 2011).

An examination of recommendations made by those who have conducted qualitative research after disasters, according to the location of that disaster, provides an indication of the difference in issues around the world. A study which looked at caring for children after Hurricane Katrina included in their recommendations to provide quality and affordable childcare for evacuees (Peek and Fothergill 2009). As outlined in previous examples, the issues for developing countries are much more likely to be about keeping children safe and recognised.

Another impact on children which seems to be largely dependent upon location is the degree of interruption to their education. Anywhere in the world, school buildings can be damaged or destroyed by winds, flooding, earthquakes or fire. The difference is in what happens next. An event with little impact on children in high-income countries and communities may have vastly more serious implications for children living in poverty (Bartlett 2008). A recurrent theme in the literature on the impacts of natural disasters on children in developing countries is the extended disruption to education, especially for girls. Children in an area of Bangladesh that floods annually in the monsoons nominated the loss of educational opportunities as one of their main concerns (Martin 2010). Schools may be swept away, or become difficult or dangerous to access, or the fees may be out of reach financially until the family is able to return to generating income. Girls are sometimes taken out of school to meet those financial obligations, while boys in their family continue to attend.

Responding to the Issues

The variety of issues outlined above which contribute to increased vulnerability of women and children to climate change impacts such as an increase in natural disasters, demands similar variety in responses. There is no one easy solution. Indeed some of the issues arising from cultural norms may especially difficult to tackle. However, there are a number of recommendations and innovative solutions, many of which use principles of crime prevention—looking at factors related to situational and opportunistic crimes.

The Gender Disaster Network provides six general principles for including gender issues in relief and reconstruction:

1. *Think big*—gender equality and risk reduction principles should guide all aspects of disaster mitigation, response and reconstruction; and utilise the small window of opportunity for change in the immediate aftermath of an event.
2. *Get the facts*—gender analysis is not optional but imperative to direct aid and plan for a full and equitable recovery.
3. *Work with grass roots women*—women's community organisations have insight, information, experience, networks and resources vital to increasing disaster resilience.
4. *Resist stereotypes*—base all initiatives on knowledge of difference and specific cultural, economic, political and sexual context, not generalisations.
5. *Take a human rights approach*—democratic and participatory initiatives serve women and girls best, and both males and females need to be assured of conditions of life needed to enjoy their fundamental human rights, as well as simply survive.
6. *Respect and develop capacities of women*—avoid overburdening women whose workloads will already increase after a disaster, identify and support women's contributions to informal early warning systems, school and home preparedness, community solidarity, socio-economic recovery and extended family care (Gender and Disasters Network 2005).

These general principles must then be translated into concrete actions if they are to make a difference. Given the number of reports of harassment, abuse and violence in IDP camps and shelters, clearly some closer attention must be paid to their design and operation. Recommendations on this have come from a variety of sources, and include avoiding crowded sleeping conditions, lack of privacy and partitions, lack of separate toilet and washing facilities for men and women, basic facilities being in secluded areas far from the main camp, and poor lighting especially near water, toilet and washing facilities (Delaney 2006). These ideas are not unfamiliar as concepts of "crime prevention by design".

A study looking at the sexual exploitation of women and children displaced by conflict situations has made recommendations and observations which will be equally applicable to disaster response given that the lived experiences are often similar (Ferris and Petz 2011). This study included interviews with victims, non-government organisations, UN agencies and governments in Southern Sudan, Ivory Coast and Haiti during 2007. Under-reporting because of fear or feelings of powerlessness, and ineffective responses by the agencies were found to be significant issues. The complex relationships and interactions between local and international agencies, combined with existing high levels of discrimination against women were found to be major contributing factors (Csáky 2008).

Increasing the involvement of women can be particularly difficult in the very communities where it is needed most—those in which cultural norms rigidly designate roles for men and for women. However, it makes sense that the people who use a particular resource should be directly involved in sustainability initiatives around that resource, since they are the ones who best know their requirements. The Hamirpur district in the North Western Indian state of Himachal Pradesh is a very conservative region where women are traditionally not allowed to speak to non-kin males, and where the collection of water is one of the duties traditionally performed by women. A non-profit public research and development organisation worked there for over 3 years to increase capacity building and get women involved in the maintenance of the resources like wells, that they need to use on a daily basis (Mehta 2009).

Children require special consideration following disasters and this will be equally applicable to situations arising from other climate change impacts such as increases in heat waves and rising sea levels. These special considerations concern not only the children themselves, but their caregivers, since the presence of a child in the household is a risk factor for adults' distress during disasters, and the most prominent risk factor for a child's distress is their parent/adult's distress (Ronan and Johnston 2005). Thus, looking after the well-being of families will be a head start in looking after the well-being of children. Adequate support of the mental health of parents and caregivers will act to reduce the risk of child abuse. The provision of somewhere to conduct schooling and safe places for children to play will not only facilitate the continuation of education, but decrease the risk of sexual exploitation and abuse of children as they spend less time unsupervised.

There are some innovative solutions that provide some positive news. In Bangladesh where large parts of the country are flooded annually in the monsoons, schooling for children is usually disrupted at that time of year. In response to this, some areas have school boats, equipped with libraries, computers and lights, all running on solar energy (Martin 2010).

As the predicted impacts of climate change take shape, and the need for an understanding of the social implications becomes clear, issues of gender and age must be firmly embedded. Disaster statistics disaggregated by gender and age group should be a minimum, and yet these are scarcely available. Better design and operation of camps and shelters for displaced persons, and a clearer recognition of the needs of men, women, youth, boys and girls is essential. Clearly location and culture are important factors in natural disaster responses, and there is limited knowledge of how climate change impacts will affect local areas and regions. We do not have control over all of the ingredients for the tragic outcomes touched upon in this chapter, and the ones related to climate change and variability are not going away in the foreseeable future. However, we do have control over many of the social factors and vulnerabilities that turn extreme weather events into disaster situations. There is much work to be done, and in some places, the need is urgent.

Men and Boys

The literature overwhelmingly represents women and girls as victims and amongst the poorest of the poor, and as disempowered by gender inequalities in their communities. During extreme events, women and girls are said to have less equitable access to resources, more likely to be subjugated by food hierarchies and less able to move around due to cultural norms that privilege men over women and boys over girls. Women are more likely to be primary carers and to be over-burdened with work due to the division of labour, especially following disaster. Simultaneously, they are nonetheless seen as primary agents of change in the event of catastrophe or crisis. There is widespread agreement in the literature that women and girls are at greater risk of dying (particularly drowning) in the immediacy of disaster; at greater risk of physical and sexual abuse in transitional accommodation such as camps, tent cities and evacuation centres, that domestic violence against women is likely to increase post-disaster and that children in certain regions are highly vulnerable to trafficking and sexual abuse during and post-disaster.

Aside from positive representations of men as protectors (of family, kin and community) and providers (still primary providers in many communities around the world), adult men and male youth are portrayed as predators and perpetrators of physical and sexual violence during and post-disaster. Adult men are also seen as agents of obstruction to women's empowerment and overall as part of the problem for women and girls rather than partners in solution. Male youth are portrayed as troublemakers and in certain contexts, as prone to form gangs and engage in gang-related violence. This tendency to "criminalise" men and boys, particularly in disaster situations, relies on generalisations. We know less about males in these same communities whose behaviour is the antithesis to this, which leaves a gap in our knowledge about the positive contributions different groups of men (e.g. elderly men, adult men, male youth and boys) are making, in the context of different types of events.

Rarely are adult males or male youth represented as victims. Boys, on the other hand, are almost invisible, except for a mention as sometimes victims of trafficking and sexual abuse, but almost always as secondary to girls. This is despite the fact that it is now widely acknowledged that boys are just as likely to be the targets of paedophiles as girls, but less likely to report sexual assault and that non-reporting is exacerbated in certain cultural contexts. Where the term "children" (read here girls *and* boys) is used, the literature invariably resorts to a discussion about the vulnerabilities and needs of girls. If boys are mentioned, it is an adjunct to girls or for comparative purposes. For instance, in specific cultural and religious contexts, boys are represented as especially privileged over girls (e.g. brothers are "saved" over sisters in disasters, receive food before their sisters, are more likely to escape floods because of tree climbing skills, or are able to move about more freely than their sisters). Like women and girls, men and boys are treated as a homogeneous undifferentiated group. This means that different groups of men are overlooked in the literature including elderly men, poor men, male youth-headed households, widowers, and "men with marginalised sexual identities who may be subject to violence or isolation, especially during periods of social crisis" (Enarson 2009).

The overarching consensus seems to be that women and girls are most disadvantaged by climate change and disaster because of pre-existing socio-economic and socio-cultural inequalities (see Brody et al. 2008, p. 2). However, "...there is a heavy reliance on generalisations, which cannot hold true for all people in all places", so "the default response has been to talk about women and *their* vulnerabilities" (Demetriades and Esplin 2008, p. 25), thus privileging the voices of women and girls in the developing world and to a lesser extent the developed world, over the voices of men and boys everywhere. The authors also note that "experiences of powerlessness can leave men and boys like women and girls vulnerable to climate change and disaster, albeit often in different ways" (Esplin with Brody 2008, p. 2, cited in Demetriades and Esplin 2008, p. 25) and caution against approaches that focus on women in isolation:

> Women live in communities, they live in families, they live *with men.* Abstracting women from their social realities eclipses the relational nature of gendered power and the interdependency of women and men, and paints a distorted picture of women's vulnerabilities, choices and possibilities. This is exacerbated by a tendency to treat women as a homogeneous undifferentiated group

The disaster literature in particular reveals much about the victimisation of women and girls by adult men and male youth, but less about the victimisation of adult men, male youth and boys in the same circumstances.

"Although both men and women are vulnerable to climate change, the causes of their vulnerability and their experiences of it are different, along with their capacity to cope and adapt" (ICIMOD 2009, p. 1). While gender relations typically empower men as decision-makers with greater control than women over key resources (Enarson 2009, p. 1), gender identities and norms also increase their risks. This "gendered vulnerability does not derive from a single factor such as household headship or poverty, but reflects historically and culturally specific patterns of relations in social institutions, culture, and personal lives" (Enarson 1998). "Gender shapes vulnerability and therefore one's capacity to adapt to change, whether it be to catastrophe or incremental change" (Alston 2011, p. 64).

Notions of masculinity place males at risk in different ways. For instance, the following gender roles create risk for men:

- Gender roles that promote risk-taking.
- Gender roles that promote invincibility.
- Gender norms related to men's reduced risk perception and increased tolerance of risk [risk-taking] can endanger men and their families.
- When preparedness and evacuation are choices, men may delay and underprepare or be injured in the clean-up period (see Enarson 2009).

Men's "invincibility" also means they are less likely to seek help or assistance, whether that be economic, social, psychological or health-related. Men and boys are also among some of society's most high-risk groups, including the homeless, refugees, poor and low income, the mentally ill, drug and alcohol-addicted and socially isolated.

Issues and social circumstances that can lead to increased vulnerability for males during and post an event include:

- Emotional stress and anxiety.
- Rising indebtedness.

- Social isolation (social ties to family and kin promote resilience but may be tenuous for some men, including widowers (Enarson 2009)).
- Displacement—male economic migrants displaced to risky living conditions (Enarson 2009).
- Out-migration—involuntary separation from intimate partners, family and kin (e.g. working away from "home" for extended periods of time). In these situations men are distanced both geographically and emotionally from their traditional support networks and therefore more prone to loneliness, anxiety and depression.
- Exploitation—men and male youth may be exploited for their labour (long hours, poor wages, hard labour).
- Emasculating experiences for men such as being unable to protect family members during extreme events, or to save strangers.
- Male youth and boys separated from their fathers at critical stages of their development.

In relation to men's poor health-seeking behaviour, Alston (2011) makes an interesting observation that women have taken on the role of guardians of men's health. In one instance, for example, gender socialisation had a profound impact on a group of young Sudanese men in a Western Ethiopian refugee camp, "who continued to starve, despite receiving food aid, because the food they were given needed to be cooked and they never learned to cook" (WHO 2002, p. 2).

It is evident that the degree and pace of change in gender roles and relations (e.g. radical, evolving) is in part determined by the type of event (flood, tsunami, drought) and pace of the event (rapid onset; slow onset) as well as intersecting variables such as poverty, age, class, caste, ethnicity, race, and so forth. Capacity to cope in the immediacy of an event and to adapt over time is also influenced by external factors such as how rapidly practical help arrives, the level of empathy or indifference from local and international aid communities, where the event occurs geographically and pre-existing stressors such as social or military conflict, racial discrimination, economic uncertainty, political instability and religious or cultural constraints that limit people's ability to make the necessary changes or whose circumstances deny them the resources, choices and autonomy to take control of their destiny.

Men's coping strategies take different forms but are typically driven by the provider stereotype (e.g. out-migration for work; getting back to work as quickly as possible), and the protector stereotype (e.g. first responder; search and rescue; protector of home and family; protector of strangers and community; doing the dirty work of cleaning up and rebuilding). These strategies place men at risk by exposing them to significant trauma, environmental toxins, and involuntary separation from social and emotional support networks. In such circumstances they are vulnerable to death, injury and wounding, physical and psychological trauma, immediate and long-term health repercussions, loneliness, depression, anxiety, and at the extreme end, suicide. In seeking to provide for their families, men who out-migrate may be subject to labour exploitation (long hours, low wages) and in the immediacy and aftermath of disasters, threats of violence from other men. Some studies suggest that "seasonal migration with longer travel distances may put men more at risk to contracting infectious diseases, notably HIV and sexually transmitted diseases" (Leduc 2009, p. 3). This statement contains an inherent value judgement (and generalisation) about men's behaviour; however, if this conversation is to be had, then it needs to be had in relation to women who migrate out too.

The literature indicates that men's perceptions and tolerance of risk differs from women's (e.g. men are risk-takers, women are risk-averse). As an illustration of gendered differences in risk perception, Delaney and Schrader (2000, p. 27) note men's and women's differing perspectives on the worst and least important impacts of the El Chile disaster in Nicaragua:

> For men the worst impact was "decreased coffee production" and the least important impact "lack of water". For women the worst impact was "fear" the least important impact "less income" "more work"

The way in which men perceive risk may result in them underestimating impending danger and this may play out in a reluctance to evacuate in a timely way, in driving or wading through rising flood water, or attempting to rescue someone else in fast-flowing water.

The impact of gender role socialisation on men and boys places them at greater risk in certain circumstances. The two predominant roles are protector and provider.

Men as Protectors

The gender division of labour also often places men at high risk in the emergency relief phase of disasters (Enarson 2009, p. 1). As first responders and protectors of the community, men are often exposed to toxic emissions, especially in the advent of natural and technological disasters. Here the men of 9/11, the men of Chernobyl and the men (predominantly) of Fukushima come to mind. This expectation places a greater responsibility on them for risk-taking during and after disasters, both within their households and as first responders or rescue workers (WHO 2002, p. 2). This can have serious consequences for men's long-term physical, mental and emotional wellbeing as well as their capacity to fulfil their role as husbands, partners, fathers and community protectors. In their role as first responders, men also face the dilemma of protector of family/community, an issue that is compounded where both men and women in a household are first-responders (see Enarson 2009)

Although it is gender stereotypes that expose men to dangerous environmental toxins, it is biology (sex) that determines how those toxins will affect their bodies in the short and long term. Impacts can include reproductive disorders, and chronic and incurable diseases. There are also wider social repercussions as a result of the stressors involved.

> During rehabilitation, whilst women maintained the household and social networks, men were involved in dangerous reconstruction efforts; some men were also taking part in increased gambling, increased consumption of alcohol, and some were displaying greater aggression (Delaney and Shrader 2000, cited in Nelson et al. 2002, p. 56)

We need to know much more about which groups of men are involved in reconstruction work and how separation from their normal emotional support networks (intimate partners, families, peers and friends contributes (or not) to the types of behaviour described above?

The negative side of the protector role is that men can feel emasculated by disempowering experiences such as failing to "save" loved ones and strangers in times of disaster. This can lead to self-blaming behaviour, feelings of inadequacy, low self-esteem, depression, and in worst-case scenarios suicide. Instances of suicide have also been recorded among male farmers in rural Australia in response to prolonged drought (see Alston 2011). Similarly, if men cannot provide adequately for their families they are likely to suffer the consequences in a way that undermines their masculinity and sense of identity.

Men as Providers

In many parts of the world, men are still the primary bread-winners for their families. A typical coping mechanism for men following disaster is to return to work as quickly as possible. They are more likely to migrate out for economic reasons to secure primary or supplementary income. If men lose their livelihoods, are unable to earn the level of income they are accustomed to or cannot carry out this role in a customary way, the consequences are likely to be far-reaching. A study by the World Health Organisation (WHO 2002, p. 2), noted that: "in the aftermath of Hurricane Andrew in the United States, men who had traditionally been the family providers (and protectors) struggled with feelings of inadequacy and failure".

Malin Jennings (cited in Parbring 2009) observes a similar effect on the lives of men in small Inuit communities in Greenland, where a warmer climate has impacted on hunting (typically a male role). Here the ice freezes later and is thinner and also melts earlier than before, making it more difficult for men to provide for their families, resulting in self-esteem problems and social conflict:

> Climate change robs Arctic men of their self-esteem. They can't feed their families by hunting. They're no longer proud of themselves.
>
> For many men, self-esteem is bound up with being the strongest and best hunter.
>
> When that knowledge is no longer in demand, they lose their sense of pride. In Greenland, men are now living off benefits instead, but they have nothing to do—so they congregate in bars—which leads to alcoholism.
>
> Hunters used to be the pinnacle of society but this is no longer the case. Their skills and experience have been rendered useless, and this is creating social problems (see Parbring 2009)

Men's poor help-seeking behaviour (reluctance to seek all kinds of help) also creates situations of humiliation for them. In the following case it was economic assistance

> …in 2001 the flood destroyed her and her husband's wet season rice crop. He asked [his wife] to borrow money from the private money lender so that he could buy some new seed stock and fertiliser. Her husband would not go because he felt embarrassed to talk to the money lender (Flood impact on women & girls in Prey Veng Province, Cambodia, CARE International 2002, cited in UNDP 2010, p. 2).

Vulnerabilities and the Social Impact of Disasters

The following examples intentionally privilege the perspectives and experiences of different groups of men, youth and boys as providers, protectors, carers, victims and agents of change.

The Widowers of Lamteungoh, Indonesia

The literature reveals that a disproportionate number of women and children die in disaster, but less is known about the men they leave behind. Enarson (2009) provides a rare example from the *Washington Post Foreign Service.*

> In the Indonesian coastal town of Lamteungoh there are 105 widowers and only 19 widows. This is because many were fishermen who survived at sea or farmers in the hills about the high waterline. "Their wives and children were killed at home not far from the beach when the driving waves turned the village into ruins on December 26, 2005. These rugged men are now grappling with unfamiliar roles, dependent on one another and uncertain about what comes next. With their families gone, some say their lives have lost purpose. They are caring for children in communal style and tending to the injured. They are struggling to move through their grief and reclaim their future" (Nakashima (2005) cited in Enarson 2009, p. 1).
>
> "Life today has no meaning at all for me". "Now suppose I find a job and make money. To whom can I distribute it?" (Baharuddin, age 49).
>
> Source: Ellen Nakashima, Washington Post Foreign Service, 25.1.05, p. A01 (cited in Enarson 2009, p. 1)

As noted by Delaney and Schrader (2000), "relations between natural disasters and gender do vary—more men died than women in Hurricane Mitch" (cited in Nelson et al. 2002, p. 55). More men die in severe-weather events in the United States (see Fothergill 1998), more men than women (100 vs. 73) died in the Black Saturday fires in Victoria (see Parkinson 2011), Australia and "in Paris, France in 2003, the heatwave-related risk increased for unmarried men, but not unmarried women" (see WHO 2010, p. 10).

The Incarcerated Male Youth of Katrina

"Treated like trash" is how a 15 year old boy described his experience of incarceration as Hurricane Katrina made landfall in New Orleans. During their evacuation these incarcerated children (some as young as 10 years old) experienced prolonged hunger and dehydration (3–5 days). En route to their new destination they were shackled together with plastic handcuffs, forced to walk through rising flood water and threatened with violence from those charged with protecting them. They witnessed the beating and maceing of others, and one child witnessed the death of a man and the disposal of his body. They were exposed to bacterial infections from rising water in cells contaminated with faeces from backed-up toilets and some were forced to drink it because they were so thirsty. Two boys were inappropriately placed in a dormitory with adult inmates.

These narratives of abuse by the incarcerated children of Katrina serve as a precautionary tale for policy makers in the event of sudden onset event. Many of these children were not convicted, awaiting processing through the criminal justice system.

> C.M., a 16-year-old boy, stated, "A few hours after the storm hit, the water started rising. That night the water started coming out of the toilet and the drains. It smelled like straight swamp water. I was crying and thinking about my people because right before the power went out we saw what was happening on the news and saw the Ninth Ward flooding. Kids were really upset because most of them were from the Lower Ninth" (JJPL Undated, p. 16)

> C.S., a 15-year-old boy, "We had human faeces floating around us in the water ... we was forced to survive in for 3 days. I still have little sores on my skin. I can't seem to get that smell out of my skin. ... [M]aybe it's all in my head but that smell will be with me, and be in my head for a very long time" (JJPL Undated, p. 17)

Many children said the passage through the waters was one of the most difficult parts of the experience.

> O.S., a 14-year-old boy, stated once they arrived at the Broad Street Bridge, the children were threatened by armed, uniformed officers whom O.S. believed were from the New Orleans Police Department. "They had big guns. ... They told us that the mayor said 'We can shoot to kill.' There was military there, too, but it was mostly NOPD. NOPD beat up an adult prisoner. They busted open his head. ... You could see the meat" (JJPL Undated, p. 19)

Sexual Violence Against Boys

Gender-based violence is typically framed in terms of violence against women and girls. Sexual violence against men and boys is largely unconsidered in the literature. It is also harder to assess the scope of sexual violence against men and boys because many are reluctant to report it due to fear of stigmatisation. Sexual violence against men and boys often goes unrecognised and unreported—in conflict situations sexual violence against men is typically an intimidation strategy, whereas young boys (like girls) are more typically abused by unscrupulous men (and women) among them the aid workers they are relying upon to help them. "The sex industry often becomes part of the interaction between the refugee or displaced population and the local community. Men and boys may also be at risk of sexual abuse in such circumstances" (WHO 2002, p. 2).

Research by the Save the Children Fund in the United Kingdom (see Csáky 2008) illustrates that "significant levels of abuse of boys and girls continues in emergencies, with much of it going unreported" [and that] "victims include orphans, children separated from their parents and families and children in families dependent on humanitarian assistance". Perpetrators of sexual and gender-based violence are sometimes the very people survivors depend upon to assist and protect them' (United Nations High Commissioner for Refugees 2003, p. 14), and include both male and female perpetrators.

> There's a man who works for [an international organisation] who gave 400 Haitian gourds to a 13 year old and he took his bottom with his two hands and then he went away with him and raped him (Young boy, Haiti quoted in Csáky 2008, p. 7)

> One day two boys who ran errands for the [international organisation] saw a woman go into the bush and give a boy of 13 a blow-job (Young boy, Cote d'Ivoire, quoted in Csáky 2008, p. 7)

Boys told researchers they were fearful of reporting the abuse for a number of reasons including "people don't report it because they are worried that the agency will stop working here and we need them" (Teenage boy, Southern Sudan quoted in Csáky 2008, p. 13); "some children are scared they might be killed by the perpetrator" (Young boy, Haiti, quoted in Csáky 2008, p. 13); "Who could we tell? We wouldn't tell the police because they are afraid of the peacekeepers and they can't do anything...anyway, I've heard that the police do this kind of abuse too' (Young boy, Haiti, quoted in Csáky 2008, p. 14). A teenage boy said, "the biggest encouragement would be to make people feel safe to report" (Csáky 2008, p. 20).

Responding to the Issues

The approaches of gender sensitivity, situational crime prevention, and change and conflict management all have something to bring to the table in terms of precautionary approaches to preventing harm to people, places and the environment. Gender-sensitivity could be the lens through which the issues are examined. Part of this entails learning more about the gendered nature of living through disaster situations (see Table 9.1). Enarson (2009, p. 4) notes how gender-sensitivity training for soldiers in East Timor was instrumental in reducing violent incidents against women. The flip side to this is how similar outcomes might be achieved to reduce male on male violence, as well as female on male violence. This requires closer attention to how gender roles are socially constructed and re-constructed in particular cultures and circumstances.

A precautionary approach might anticipate the unexpected; predict the triggers for gender conflict; initiate advance training of workplace and school alert teams in

Table 9.1 Gender experiences in disaster situations

Circumstance	Men (examples)	Women (examples)
Vulnerability	Search and rescue	Forced to stay at home
	First responder role	Risk lives for children
Risk perception	Ignore emergency warnings	Forced to evacuate
	Last to evacuate	Fear of rogue men
Coping strategies	Temporary migration	Resettlement camp
	Less assistance	Greater family responsibility
Needs and priorities	Return to work as soon as possible	Fear for loved ones
	Dealing with indebtedness	Food and shelter
Social composition	Men seen as mobile	Female-headed households
	Living alone	Enforced communal living
Creation of new vulnerabilities	Sexual/physical violence	Sexual/physical violence
	Drug/alcohol abuse	Anxiety over relationships
New gender roles	Domestic duties	Engage in traditional male work roles
	Widowed men	Head of household
Gender relationships	Reliance on outside aid	Women's programs/male exclusion
	Men seen as perpetrators	Focus on women's empowerment

Source: Drawing and modelled on Delaney and Shrader (2000, p. 14)

first aid and evacuation; consider gender-balanced evacuation and health teams; raise awareness about the reproduction of stereotypical images and narratives in the mass media, promote the formation of gender-balanced policing teams to take reports of sexual assault; or urge Environmental Protection Agencies to work with industry to storm-proof industrial precincts and mine sites (subject to inundation during extreme flooding).

A situational and community crime prevention approach is inherently precautionary, in that it seeks to alter the built environment to prevent crimes of opportunity (e.g. anticipate the types of crimes likely to occur during different phases of an event). In this context, precautionary approaches might include community "policing" of water and fuel-wood routes; lighting and "policing" of public toilets in evacuation centres; portable privacy screens in evacuation centres and transitional communal shelters; improved planning and layout of "tent cities"—e.g. in "streets", interspersed with larger "community" tents to accommodate law enforcement, health-care, and other social services as well as providing separate space for communities to gather or the formation of gender-balanced 'neighbourhood watch groups' in tent cities and evacuation centres could provide a deterrent to crime and criminality.

A horizon scanning approach (see White and Heckenberg 2011) could be applied to examine different types of events (e.g. climate change, extreme-weather, disaster) and the pace of those events (e.g. slow onset, rapid onset); to forecast changes in gender roles and relations, to anticipate potential conflict and to formulate a range of strategies for conflict resolution In terms of events, horizon scanning might be used as a tool to imagine the unimaginable (Fukushima being an example of an event previously unimagined), as well as to forecast how different types of events might disrupt or change gender roles and relations and the wider implications of those changes for gender relations including intimate partners, families, communities, regions and nations, within specific contexts. In this regard Haq et al. (2008, p. 2) make a distinction between current and future climate change, pointing to "the need to reduce and adapt to risk on both timescales—the mitigation of future changes in climate is a long-term task, both intergenerational and international in its nature".

Conclusion

It is important to avoid assumptions about how people will adapt to environmental change, including climate change, and the consequences of this for gender relations (Nelson et al. 2002, p. 57). Australia, for instance, has a diverse ecological landscape and a multicultural population; however, it is given that the nation will pull together in a crisis, with men and women working alongside one another in the clean-up (e.g. the 2010 Brisbane floods). However, it is much more difficult to navigate the diverse social and environmental landscape of a place like the Himalayas for instance, where people are separated not only by different gender roles, expectations and practices but also by language, culture and religion. There is a need to better understand the often complex terrain upon which these life-changing events play out and to tailor interventions to take account of the diversity of geographical, economic, political, social, cultural, religious and gender landscapes across different regions of the developed and developing world (see Leduc 2009).

Although climate change is a global phenomenon its effects are experienced most acutely at the local level, as are the impacts of extreme-weather events and disaster. Therefore the solutions must respond to the everyday lives of people, in the context of their relationships with one another, and the economic, political, social, religious and cultural realities of the places in which they live, work, learn and play. Gender *is* relational—we live in families and communities. It is only with the full and equal participation of women and men, male and female youth, and boys and girls that we will build safer, more just and resilient communities. We need a repository of gender knowledge that is context-specific to location and type of event as well as respectful of the diverse gender landscapes of different communities of people.

References

Action Aid Nepal. (2007). *Violence Against Women In Disaster Situation: Women's Report Prepared in the Post Flood Context, Mahottari, Dang and Banke Districts of Nepal.* Kathmandu: Action Aid Nepal.

Alston, M. (2011). Gender and climate change in Australia. Journal of Sociology. Vol 53, 11 October, 2010. Sage Publications on behalf of the Australian Sociological Association. DOI: 10.1177/1440783310376848. http://jos.sagepub.com/content/47/1/53.full.pdf.

Ariyabandu, M. M. (2009). Sex, gender and gender relations in disasters. In E. Enarson & P. G. D. Chakrabarti (Eds.), *Women, Gender and Disaster: Global Issues and Initiatives* (pp. 5–17). New Delhi: Sage Publications India Pty Ltd.

Arts, K. (2009). A child rights perspective on climate change. In M. A. M. Salih (Ed.), *Climate Change and Sustainable Development: New Challenges for Poverty Reduction* (pp. 79–93). Cheltenham: Edward Elgar Publishing Limited.

Bartlett, S. (2008). *Climate Change and Urban Children: Impacts and Implications for Adaptation in Low- and Middle-Income Countries (Theme: Climate change and Cities - 2)*. London: IIED.

Brody, A., Demetriades, J., & Esplen, E. (2008). *Gender and Climate Change; mapping the linkages: A scoping study on knowledge and gaps*. BRIDGE, Institute of Development Studies (IDS), University of Sussex, United Kingdom. http://www.bridge.ids.ac.uk/reports/Climate_Change_DFID.pdf.

Chew, L., & Ramdas, K. N. (2005). *Caught in the Storm: The Impact of Natural Disasters on Women*. San Francisco, CA: The Global Fund for Women.

Cottrell, A. (2008). Quiet achievers: Women's resilience to a seasonal hazard. In K. Gow & D. Paton (Eds.), *Phoenix of Natural Disasters: Community Resilience* (pp. 181–193). New York: Nova Science Publishers Inc.

Csáky, C. (2008). *No one to turn to: The under-reporting of child sexual exploitation and abuse by aid workers and peacekeepers*. London: Save the Children Fund.

Delaney, S. (2006). *Protecting children from sexual exploitation and sexual violence in disaster and emergency situations*. Bangkok: EPCAT International.

Delaney, P., & Shrader, E. (2000). *Gender and Post-Disaster Reconstruction: The case of Hurricane Mitch in Honduras and Nicuragua: Decision Review Draft*. The World Bank: LCSPG/LAC Gender Team.

Demetriades, J., & Esplin, E. (2008). The gender dimensions of poverty and climate change adaptation. *Institute of Development Studies, 39*(4), 24–31.

Gender and Disasters Network. (2005). Gender equality in disasters: 6 principles for engendered relief and reconstruction. www.gdnonline.org/resources/genderbroadsheet.doc. Accessed 4 Sep 2011.

Enarson, E. (1998). 'Through women's eyes: A gendered research agenda for disaster social science. *Disasters, 22*(2), 157–173. http://asci.researchhub.ssrc.org/through-womens-eyes-a-gendered-research-agenda-for-disaster-social-science/attachment. Accessed December 2011.

Enarson, E. (2009). Women, gender and disaster: Men & masculinities. Gender Note No. 3. Gender Disaster Network. http://www.mensresourcesinternational.org/documents/GDN_GenderNote3.pdf. Accessed October 2011.

Ferris, E., & Petz, D. (2011). *A year of living dangerously: A review of natural disasters in 2010*. London: The Brookings Institution - London School of Economics Project on Internal Displacement.

Fisher, S. (2010). Violence against women and natural disasters: Findings from post-tsunami Sri Lanka. *Violence Against Women, 16*(8), 902–918.

Fothergill, A. (1998). The neglect of gender in disaster work: An overview of the literature. In E. Enarson & B. H. Morrow (Eds.), *The Gendered Terrain of Disaster: Through Women's Eyes*. Westport, CT: Praeger Publishers.

Hannan, C. (2009). Mainstreaming gender perspectives in natural disaster management. *Regional Development Dialogue, 30*(1), 1–11.

Haq, G., Whitelegg, J., & Kohler, M. (2008). *Growing Old in a Changing Climate: Meeting the Challenges of an Ageing Population and Climate Change*. Stockholm Environment Institute, Kräftriket 2B, Stockholm, Sweden.

Hepburn, S., & Simon, R. J. (2010). Hidden in plain sight: Human trafficking in the United States. *Gender Issues, 27*, 1–26.

Houghton, R. (2009). 'Everything became a struggle, absolute struggle' Post-flood increases in domestic violence in New Zealand. In E. Enarson & P. G. D. Chakrabarti (Eds.), *Women, Gender and Disaster: Global Issues and Initiatives* (pp. 99–111). New Delhi: Sage Publications India Pvt Ltd.

Juvenile Justice Project of Louisiana, JJPL (undated). *Treated Like Trash*. Juvenile Detention in New Orleans Before, During and After Hurricane Katrina. http://jjpl.org/site/wp-content/uploads/2011/07/treated-like-trash.pdf. Accessed Dec 2011.

Leduc, B. (2009). *Gender and Climate Change in the Himalayas.* Background paper for the e-discussion on Climate Change in the Himalayas : The Gender Perspective, organised by the International Centre for Integrated Mountain Development (ICIMOD) and APMN. 5-25 October 2009. www.icimod.org/resource.php?id=257. Accessed Dec 2011.
Martin, M.-L. (2010). Child participation in disaster risk reduction: The case of flood-affected children in Bangladesh. *Third World Quarterly, 31*(8), 1357–1375.
Masai, R., Kuzunishi, L., & Kondo, T. (2009). Women in the Great Hanshin earthquake. In E. Enarson & P. G. D. Chakrabarti (Eds.), *Women, gender and disaster: Global issues and initiatives* (pp. 131–141). New Delhi: Sage Publications India Pvt Ltd.
Mehta, M. (2009). Reducing disaster risk through community resilience in the Himalayas. In E. Enarson & P. G. D. Chakrabarti (Eds.), *Women, Gender and Disaster: Global Issues and Initiatives* (pp. 57–74). New Delhi: Sage Publications India Pvt Ltd.
Mutter, J. C., & Barnard, K. M. (2010). Climate change, evolution of disasters and inequality. In S. Humphreys (Ed.), *Human Rights and Climate Change* (pp. 272–296). Cambridge: Cambridge University Press.
Nakashima, E. (2005). 'For Men of Seaside Village, Lonely and Unfamiliar Roles : Tsunami Took Heavier Toll on Aceh's Women'. Washington Post Soreign Service. 25 January 2005, Page A01. http://www.washingtonpost.com/wpdyn/articles/A33734-2005Jan24.html. Accessed 30 April 2012.
Nelson, V., Meadows, K., Cannon, T., Morton, J., & Martin, A. (2002). Uncertain predictions, invisible impacts, and the need to mainstream gender in climate change adaptations. *Gender and Development, 10*(2), 51–59.
OHCHR. (2011). *Protecting the human rights of internally displaced persons in natural disasters: Challenges in the Pacific.* Suva: Office of the High Commissioner for Human Rights Regional Office for the Pacific.
Olson, R. S., & Gawronski, V. T. (2010). From disaster event to political crisis: A "5 C+A" framework for analysis. *International Studies Perspectives, 11*, 205–221.
Parbring, B. (2009). 'Gender impact of climate change'. *Norden.* Official Co-operation in the Nordic Countries. The Nordic Council of Ministers. http://www.norden.org/en/news-and-events/news/gender-impact-of-climate-change. Accessed 10 Aug 2011.
Parkinson, D. (2011). *Gender Disaster and Violence : Literature Review.* Women's Health Goulburn North East. http://www.whealth.com.au/documents/environmentaljustice/women-disaster-violence-lit-review.pdf. Accessed Dec 2011.
Peek, L., & Fothergill, A. (2009). Parenting in the wake of disaster: Mothers and fathers respond to Hurricane Katrina. In E. Enarson & P. G. D. Chakrabarti (Eds.), *Women, Gender and Disasters: Global Issues and Initiatives* (pp. 112–130). New Delhi: Sage Publications India Pvt Ltd.
Ronan, K. R., & Johnston, D. M. (2005). *Promoting community resilience in disasters: The role for schools, youth and families.* New York: Springer.
Saito, Y. (2009). Gender mainstreaming into community-based disaster management in the context of regional development. *Regional Development Dialogue, 30*(1), 37–46.
Saroor, S. (2009). *Impact of the tsunami in the East—Muslim women's perspective.* Colombo: Social Scientists' Association.
Sayeed, A. T. (2009). Victims of earthquake and patriarchy: The 2005 Pakistan earthquake. In E. Enarson & P. G. D. Chakrabarti (Eds.), *Women, Gender and Disasters: Global Issues and Initiatives* (pp. 142–151). New Delhi: Sage Publications India Pvt Ltd.
Terranova, A. M., Boxer, P., & Morris, A. S. (2009). Changes in children's peer interactions following a natural disaster: How predisaster bullying and victimization rates changed following Hurricane Katrina. *Psychology in the Schools, 46*(4), 333–347. doi:10.1002/pits.v46:4.
Terry, G. (2009). No climate justice without gender justice: An overview of the issues. *Gender and Development, 17*(1), 5–18.
Thornton, W. E., & Voigt, L. (2007). Disaster rape: Vulnerability of women to sexual assaults during Hurricane Katrina. *Journal of Public Management and Social Policy, 13*(2), 23–49.
Todres, J. (2011). A child rights-based approach to reconstruction in Haiti. *Intercultural Human Rights Law Review, 6*, 43–85.

United Nations High Commissioner for Refugees (2003). *Sexual and Gender-Based Violence against Refugees, Returnees and Internally Displaced Persons*. Guidelines for Prevention and Response. May 2003. http://www.unicef.org/emerg/files/gl_sgbv03.pdf.

United Nations Development Program, UNDP. (2010). *Gender and Disasters.* Bureau for Crisis Prevention and Recovery. October 2010. http://www.undp.org/cpr/documents/disaster/7Disaster%20Risk%20Reduction%20-%20Gender.pdf. Accessed Dec 2011.

White, R., & Heckenberg, D. (2011). Environmental horizon scanning and criminological theory and practice. *European Journal on Criminal Policy and Research, 17*, 87–100.

World Health Organisation, WHO. (2002). *Gender and Health in Disasters.* World Health Organisation, Department of Gender and Women's Health, Geneva, Switzerland. http://www.who.int/gender/other_health/en/genderdisasters.pdf. Accessed Sep 2011.

World Health Organisation, WHO. (2010). *Gender, Climate Change and Health.* Public Health and Environment Department, Health Security and Environment Cluster, World Health Organisation, Geneva Switzerland. http://www.who.int/globalchange/GenderClimateChangeHealthfinal.pdf. Accessed Dec 2011.

Chapter 10
Natural Disasters and Crime: Criminological Lessons from Hurricane Katrina

Caroline C. Nobo and Rebecca D. Pfeffer

Introduction

Perhaps one of the most severe consequences of global climate change is an increase in the frequency and intensity of extreme weather events. An unanticipated natural disaster can devastate an area with physical damages; and from a criminological perspective, it presents law enforcement with an unprecedented moral and organizational challenge. Due to the likelihood of natural disaster intensification in coming years, public safety officials need to better understand how the consequences of an unstable natural environment can affect delinquency and policing. Maintaining low crime rates is a difficult enough task without added pressures, and at-risk communities are in a position to be pushed over the edge by such extreme events. Better communication between the climate science and criminal justice communities will help to reduce the decay of public safety agencies following a natural disaster. On all fronts, the uncertain future between law enforcement, crime, and natural disasters must be addressed.

In the days following the landfall of Hurricane Katrina, an international audience was held captive by the media coverage of individual tragedies nested within the incomprehensible greater event of a natural disaster. Although the US Gulf Coast region suffered widespread damage, the catastrophic levee failure of New Orleans left the city in particularly grave circumstances. Media portrayal of the hurricane vacillated wildly between tales of order—heroism, selflessness, and good Samaritans, and those of chaos—looting, violence, and desperation. Only recently we have begun to understand the reality of the initial hours, days, and

C.C. Nobo (✉)
Charlestown, MA 02129, USA
e-mail: cc.nobo@gmail.com

R.D. Pfeffer
Boston, MA 02115, USA
e-mail: rebecca.pfeffer@gmail.com

R. White (ed.), Climate Change from a Criminological Perspective,
DOI 10.1007/978-1-4614-3640-9_10, © Springer Science+Business Media New York 2012

months following the impact of Katrina's landfall on 29 August 2005, which is that area residents, local and federal government agencies, and local law enforcement were all grossly unprepared.

This chapter discusses the criminological lessons of Hurricane Katrina in two main parts: (1) the perceived and actual looting of Katrina victims and (2) the malfeasance of law enforcement officers post-hurricane and its implications for the New Orleans Police Department. We explore these subjects through the analysis of natural disaster social reaction literature and sociological theories on social disorganization and social ecology. Hurricane Katrina is used as a case study to examine the reactions to a natural disaster by a seemingly well-prepared and well-informed government; with observations of both victims and law enforcement. Vast literature exists on the social response to natural disasters and the large-role the media plays in the misrepresentation of the event. In addition, multiple articles and government reports published after the storm address the law enforcement lessons learned from Hurricane Katrina. This chapter states that while there are necessary preparations to reduce chaos and crime following a disaster, the inherent structural disorganization following such an event is mostly unavoidable.

Upon reflection 6 years after the storm, the individual events of looting or civil unrest seem petty and less important. The lasting legacy of disorder sits squarely with law enforcement institutions, which we now understand failed at maintaining an air of authority in the city. This failure had severe consequences for the people of New Orleans and the surrounding region, perhaps most importantly a lasting loss of faith in the criminal justice system.

Background

In August 2005, Hurricane Katrina raced across the Atlantic Ocean and made contact with the US Gulf Coast. Escalating at times to a Category 5 hurricane, the storm displaced hundreds of thousands of area residents, resulted in the deaths of over 1,500 individuals, and is estimated to have caused more than $200 billion in damages, making it one of the costliest storms ever to affect the United States (Congleton 2006). The storm also had significant social effects on residents of the Gulf Coast as it resulted in widespread displacement from the region (Varano et al. 2010; Gabe et al. 2005); separation from family (Weaver 2005), schools, jobs, and other social institutions; disruption of social networks; and sustained mental health issues for its victims (Roberts et al. 2010). It is significant to note that Hurricane Katrina disproportionately affected poor communities composed mostly of African-American citizens (Gabe et al. 2005).

As presented elsewhere in this volume, it has been established that climate change has and will continue to be accompanied by changes in the intensity, duration, and frequency of weather and climate extreme events that are unusual in occurrence and have massive destruction potential (US Climate Change Science Program 2008). While even in a stable climate a small number of natural disasters caused by weather extremes are to be expected, human-induced warming is known to affect climate variables such as temperature and precipitation. Small changes across many variables will result in larger changes, resulting in a more frequent occurrence of natural disasters (CCSP 2008).

Hurricanes and flash floods rank as the most lethal and expensive natural catastrophes (National Oceanic and Atmospheric Administration 2011), both of which hit New Orleans in the summer of 2005. Observed changes in hurricanes as measured by the power dissipation index (PDI), which evaluates storm intensity, duration, and frequency, show that hurricanes in the North American Atlantic Ocean have substantially increased in storm destructive potential since 1970. According to a 2008 report produced by the US Climate Change Science Program there will likely be more frequent low-pressure systems outside of the normal North American region, with wind strength and wave heights on par with those of Hurricane Katrina. Outside of the tropics, storm tracks are shifting northward and the strongest storms are becoming even more severe. Hurricane Katrina effected lasting devastation on the economy, population, and industries of the Gulf Coast. Such a grand effect on a first-world country with resources for disaster planning and management means that the potential for destruction and loss of life is extraordinary in other developing nations. While exact projections are impossible, we know that high-income countries measure approximately 500 deaths per disaster, whereas low-income countries suffer approximately six times that amount: 3,000 deaths per disaster (Lindell and Prater 2003).

Based on historical incidents and multiple projections by climatologists, federal, state, and local governments have recognized the need to plan for the protection of communities from natural disasters on various levels, including the training of law enforcement personnel. Evaluation of existing training reveals that it often comes too late or has gaps in the practical functionality, as was the experience with Hurricane Katrina (Rojek and Smith 2007), and as will be discussed later in this chapter. While these trainings will certainly lessen the impacts of a natural disaster, the ultimate challenge is combating the social disorganization followed by such an event. An essential first step is to enhance communication among affected agencies. The climate science community and environmental law enforcement agencies (i.e., the Environmental Protection Agency) should actively collaborate with the public safety executives who make decisions about training, risk management procedures, and natural disaster preparation.

Key Issues

Looting

Immediately following the failure of the Lake Pontchartrain levees, post-Katrina New Orleans became a media magnet. The US government's response to the disaster was widely criticized as being delayed, racially biased, and insufficient. In the aftermath of a natural disaster, the distribution of food, water, and utilities can be interrupted for an indeterminate length of time. The scarcity of federally provided resources led to prompt scavenging by the citizens of New Orleans. Reports quickly aired of rampant looting throughout the city. Framed and typified by the American media, the looting associated with Hurricane Katrina became a story as big as the catastrophe itself.

In retrospect, we understand the looting that occurred in the Katrina aftermath was primarily for basic resources that were under-supplied by government agencies tasked with providing aid. Literature on natural disasters and the myths presented by the media suggest that the looting in the aftermath of the hurricane was blown out of proportion by the repeated airing and printing of a few select images (Tierney et al. 2006). At issue are also the types of products sought after in those hectic days; today, a Google image search for "looting and Katrina" returns hundreds of images of people carrying food and water, but only a handful of pictures of men carrying beer, shoes, and clothing. The latter, of course, are the images we remember that were repeatedly used by major US news sources.

In the wake of a disaster, it is important to distinguish between criminal looting and a behavior known as appropriating. Criminal looting has been defined as "both grand and petty larceny of personal property during and after disaster impact" (Gray and Wilson 1984). Appropriating behavior is defined as taking property for emergency purposes, sometimes, depending on the item, with the intent of returning it at a later date (Barsky et al. 2006). A study conducted by researchers who traveled through the region and conducted interviews just a few weeks after the storm concluded that there was no clear definition that distinguished looting from appropriating behavior, particularly in the minds of responders. What some police officers considered criminal and locked people up for, other officers considered understandable (Barsky et al. 2006). In general, activities that were once seen as looting, especially before the enormity of the storm's impact was understood, were later redefined as strategies of survival (Potter 2007). The ambiguity between criminal looting and appropriating behavior and their differing enforcement could not have been positive for the image of New Orleans law enforcement.

In the 6 years since the storm, numerous criminologists have examined the sensationalizing of the post-storm looting phenomenon (Barsky 2006; Dynes and Rodriguez 2007; Miller, 2006). Through interviews with law enforcement officers and victims they have shown that the frivolous looting of luxury items after Katrina was largely overstated. Crime statistics of petty larceny and robbery have also been examined and support this qualitative conclusion (New Orleans Police Department 2004–2006). It is important to understand the phenomenon following natural disasters that enables the media to portray victims as lawless and dangerous. The misrepresentation of disaster victim behavior creates a question for criminologists: what effect does the *perceived* looting and lawlessness among natural disaster victims have on law enforcement? For example, incorrect assumptions about the potential for looting can lead to misallocations of public safety resources that could be put to better use in assistance with victims (Fischer 1998).

When NOPD Superintendent Warren Riley stepped up as the head of the police department in the days following Hurricane Katrina, he was faced with bringing order to a city categorized as violent, out of control, and at risk of great tragedy. Only days before Riley's appointment, in an interview on ABC's *Good Morning America*, President George W. Bush said, "There ought to be zero tolerance of people breaking the law during an emergency such as this, whether it be looting or price gouging at the pump or taking advantage of charitable giving or insurance fraud" (Fox News 2005). Not only was this a severe doctrine for the freshly appointed Riley to uphold but also President Bush's publicly announced attitude toward crime in New Orleans sent a message to the entire nation that the city was violent and the victims dangerous. Dramatized fears of looters would later be used by defense lawyers as justification for several NOPD officers' shootings of disaster victims. In particular, a controversial declaration by Chief Riley to "take back the city" from looters; which several NOPD officers claim was interpreted as an authorization to use deadly force in previously unapproved situations. Riley vehemently denies ever stating such.

By the fifth day after the hurricane, 58,000 National Guard and federal troops were mobilized in the New Orleans region—more than three times the number deployed to any previous natural disaster (National Guard Bureau 2006). While the majority of these forces were used for search and rescue missions, the National Guard did join the New Orleans Police Department in their mission to combat looting. National Guardsmen were seen patrolling the city after the storm with weapons at the ready. To be fair, the Guardsmen were asked to support the New Orleans Police Department because a large proportion of the police force had abandoned their posts to join and assist their families in evacuation. Former NOPD superintendent Eddie Compass estimated that as many as 500 NOPD officers, or one-third of the force, fled. However necessary or well-intentioned it was, the militarization of law enforcement in New Orleans post-Katrina increased the perceived levels of crime and danger there.

Despite the focus on looting and criminality among victims immediately after the storm, the current stories that continue to conjure emotion, anger, and attention are ones of institutional disorganization and misconduct by New Orleans law enforcement.

Misconduct by Law Enforcement

Within the context of Katrina lies a deeper criminological issue, that of law enforcement malfeasance. Several New Orleans Police Department (NOPD) officers have been indicted on charges of manslaughter, obstruction of justice, conspiracy, and civil rights violations for their actions in the days following the storm. While NOPDs' lack of training on policing a post-disaster environment aided in their poor judgment (Rojek and Smith 2007), there is a larger discussion to be had about their criminal actions. It is important to understand how the psychological and sociological mechanisms of a natural disaster may lead law enforcement to participate in deviant behavior. In conjunction with this misconduct is how the actions of a city's police force following a natural disaster have long-lasting effects on the perception and functionality of that police department.

On 5 September 2005, police officers from the New Orleans Police Department (NOPD) responded to reports of gunfire on the city's Danziger Bridge. In two separate incidents just minutes apart, officers shot at an unarmed family, missing a 14-year-old boy, injuring three other family members, and killing 17-year-old James Brissette. Minutes later, officers shot at brothers Lance and Ronald Madison, killing Ronald, a 40-year-old mentally impaired man who still lived at home with his mother. After shooting Ronald six times in the back, Officer Robert Faulcon allegedly stomped on his body several times before he died. Although officers initially worked hard to cover up this story, a news crew that found itself in the right place at the right time caught much of the chaos on tape. Charges against several officers were dropped by a state court but the cases were picked up by federal authorities; on 5 August 2011, a federal jury convicted five NOPD officers on 25 charges of unauthorized shootings, the killings of two unarmed individuals, and obstructing justice by working extensively to cover up the events of that post-storm day. On 4 April 2012, a federal judge sentenced Officer Robert Faulcon to a 65-year term of imprisonment for his role in the slayings. In addition, Kenneth Bowen and Robert Gecivius were both sentenced to 40 years and Robert Villvaso to 38 years. Former detective Arthur Kaufman was given a sentence of six years for his role in attempting to cover up the misconduct that occurred. Media coverage of the shooting depicted it as a bizarre event in law enforcement history. However, bizarre it may have been, it was not an isolated incident.

In March 2011, former NOPD officer David Warren was sentenced for the murder of Henry Glover. Five days after the hurricane, Officer Warren shot Glover from a second floor lookout as Glover appeared to approach a strip mall in a possible act of looting. Good Samaritans discovered the injured Glover and took him to a makeshift police station. Upon arrival, they were assaulted by NOPD officer and placed in handcuffs; meanwhile, the injured Glover died in the backseat of their car. NOPD officers helped to conceal the homicide and one aided in the burning of the victim's body. This gross manipulation of power and act of deviance in the wake of a major disaster raises questions regarding the mental state of the New Orleans Police Department officers following the hurricane. It is impossible to know if there was the presence of malice during the shootings, but we can strive for an increased understanding of how organizational norms define the processes by which officers develop standards of acceptable occupational behavior and use of force following a natural disaster (Kane 2002).

Variations in social ecological conditions may influence police misconduct (Kane 2002), and although very little is known about how social disorganization may induce deviance among officers assigned to such communities we can theorize on its role in Hurricane Katrina. The social disorganization following Katrina created a context for police misconduct in two ways: (1) hurricane victims did not have in place the social networks necessary to organize against police malfeasance; and (2) the community was already characterized as experiencing high levels of police-citizen conflict due to preexisting NOPD deficiencies and corruption (Kane 2002; USDOJ 2011).

The same social disorganization that leads to deviance among the general public may also create contexts for police misconduct during a time of natural disaster. For example, the disruption in the moral order of a community is often a predictor of increased crime (Shaw and McKay 1942). Low socio-economic status, structural decay, and racial heterogeneity, all of which were present in many parts of New Orleans pre-Katrina, are standard characteristics of a community experiencing social disorganization. While vast literature points to victims banding together in disaster scenarios (Quarantelli and Dynes 1972, Quarantelli 1987, 2006), there is also evidence that when government institutions, including public safety agencies, are rendered almost entirely ineffective the reactions of the victims become less typical of a natural disaster.

Soon after Hurricane Katrina retreated from the gulf coast, the local police officers who had maintained their presence in the affected areas were commanded to give up search and rescue missions and other humanitarian work and redirect their focus on crime control. New Orleans became an increasingly militaristic environment and the police force was not adequately trained or prepared to do their jobs in such a setting. It is important to note that the vast majority of NOPD officers, who chose not to abandon their posts, were heroic in the aftermath of the storm. It is wrong to have superhuman expectations of these people. In the end, law enforcement officers and other criminal justice agents are victims themselves who are coping with a natural disaster. They are tasked with upholding their oath to protect and serve while simultaneously sheltering and providing for their families. We must establish strong supports for these individuals who are to serve as pillars in a community as well as cope with a traumatic event.

Among the severe consequences of the New Orleans Police Department misconduct following Katrina is the lasting legacy of the NOPD as a severely damaged and deficient public safety institution. The structural decay of their police force over the past 6 years was no doubt propelled by a lack of faith in the NOPD by its citizens. Recent reports on the current state of the NOPD are not flattering. A comprehensive investigation and scathing report issued by the US Department of Justice in 2011 found a pattern of excessive force, unconstitutional stops, searches, arrests, and use of biased-based policing (USDOJ 2011). While there is no way to provide a true causal link between the events of Hurricane Katrina and the current state of the New Orleans police force, it is clear that what happened in August of 2005 has detrimentally affected its functionality.

The New Orleans Police Department is still working to repair its seriously tarnished image. The city's police superintendent, Ronal Serpas, who took over the department in May of 2010, said he was troubled by what has come to light since the storm.

> We have to confront this and look at it head on. There have been far too many examples of men who have worn this badge and admitted in court to behavior that is an absolute insult to this city and to the men and women of this department who wear this badge with dignity and pride (Lee 2010, Ronal Serpas, as quoted in the New York Times 2010)

Try he might to make-over their image, Serpas' endeavor is repeatedly thwarted; current investigative research on the criminal justice system in New Orleans continues to produce new accounts of malfeasance in the wake of the disaster. This research includes two reports completed by the American Civil Liberties Union (ACLU), which document the terrible conditions and dangerous lack of planning at the Orleans Parish Prison during and after the Hurricane (ACLU 2006, 2007).

After Katrina, significant Department of Justice efforts were made to rebuild the criminal justice system in Louisiana. As of 2007, the DOJ had made available over $86 million in justice assistance grants and law enforcement infrastructure funds to the state. The US Attorney's Office placed an additional 13 temporary attorneys in the region to prosecute Katrina-related cases involving fraud and violent crime.

A Hurricane Katrina Fraud Task Force was also created to focus on specific fraudulent schemes surrounding the disaster. The instances of fraud that the Task Force focused on were perpetrated by both victims of Hurricane Katrina and those whose lives were unaffected by the storm but took opportunistic advantage of a broken city and system. Examples of such fraud included the establishment of fraudulent charities in which individuals falsely advertised themselves as agents of a legitimate Katrina charity; the identify theft of hurricane victims to receive monetary benefits; insurance fraud by hurricane victims submitting false or inflated insurance claims; and government benefit fraud, in which the hurricane victims filed false applications seeking benefits to which they were not entitled. These types of fraud are common surrounding natural disasters, and were even seen following the twin earthquake and tsunami that hit Japan in March 2011. Within 3 days of the catastrophe, dozens of online scammers were setting up false charities and hacking legitimate ones (Keizer 2011).

In addition to the task force targeting fraud, the Southeast Louisiana Criminal Justice Recovery Task Force was also created, tasked with training NOPD officers on better ways to harmonize post-disaster policing between local and federal government. Since August of 2006, federal law enforcement presence in the region has increased and the FBI created stand-alone "Katrina Squads" dedicated to combating the public corruption and government fraud crime problem. Additionally, the US Marshals Service created the Gulf Coast Regional Fugitive Task Force, which assists in locating and arresting individuals wanted on fraud charges stemming from Hurricane Katrina (USDOJ 2007).

Future Directions

The criminology of natural disasters is a relatively new focus and thus discussions can be vague about the precise nature of its consequences (Abbott 2008). While analysis from an academic community may be more in-depth, it often lacks the necessary policy implications. Moving forward, we need to facilitate a discussion within the government that focuses on the issue of public safety infrastructure during a disaster scenario; and how the decay of law enforcement in the wake of a catastrophe has long-lasting effects on the functionality of its police force. In particular, we should strive to understand how the psychological and sociological mechanisms of a natural disaster may lead law enforcement to participate in deviant behavior.

As well, there should be public awareness campaigns about the world's changing climate and the effect of extreme weather events on their region. If more people were aware of the threats brought upon by increased stormality, their involvement with relief organizations would increase; for example, obtaining certification as a trained disaster volunteer. The Red Cross offers courses free-of-charge that train participants to prevent, prepare for, and cope with disaster-related emergencies. A community who is prepared for such events can provide much needed moral and organizational support for law enforcement in the stricken region.

Certain policing techniques may help to bridge the gap between citizens and police officers during a disaster. For example, Japan's police force operates on a strict community policing design. The lack of looting and emergence of cultural cohesion following the Japanese earthquake and tsunami in March 2011 was largely credited to the relationship between Japanese police and their community. A direct link is maintained with the citizens through individual polices stands (kobans), most of which are within walking distance of another. Japan's visible police force has proven successful, and surveys showed that 95% of residents knew where the nearest koban was, and 14% knew the name of the officer who worked there. The positive police–community relationship that existed in Japan prior to the earthquake enabled them to be well organized and effective following the catastrophe. Specifically, preexisting positive relationships between police and citizens allowed them to see one other as victims of the disaster—rather than a threat (e.g., as a looter, criminal).

As a result of global climate change, the frequency and intensity of storms like Hurricane Katrina are likely to increase. Criminologists must work with law enforcement institutions to make it clear the imminent threat of natural disasters on the public safety community. While some law enforcement agencies have adopted pre-disaster policing plans, it is an oddity rather than a certainty. Continuing skepticism on the reality of climate change blocks progress of this movement. Thus, it is increasingly important that federal policy mandate natural disaster preparedness for all law enforcement agencies and that research on post-disaster environments continues to explore the lasting effects on criminal justice.

References

Abbott, C. (2008). An uncertain future: Law enforcement, national security and climate change. FRIDE, Oxford Research Group.

American Civil Liberties Union. (2006). *Abandoned and abused: Orleans parish prisoners in the wake of hurricane katrina*. New York: ACLU National Office.

American Civil Liberties Union. (2007). *Broken promises: Two years after katrina*. New York: ACLU National Office.

Barsky, L. E. (2006). Disaster realities following Hurricane Katrina: Revisiting the looting myth. In The Natural Hazards Center (Ed.), Learning from catastrophe: Quick response research in the wake of Hurricane Katrina. Boulder, CO: Institute of Behavioral Science, University of Colorado at Boulder.

Barsky, L., Trainor, J., & Torres, M. (2006). *Disaster realities in the aftermath of Hurricane Katrina: Revisiting the looting myth*. Boulder, CO: National Hazards Center.

Congleton, R. D. (2006). The story of Katrina: New Orleans and the political economy of catastrophe. *Public Choice, 127*, 5–30.

Dynes, R. R., & Rodriguez, H. (2007). Finding and framing Katrina: The social construction of disaster. In D. L. Brunsma, D. Overfelt, & J. S. Picou, J.S (Eds.), The Sociology of Katrina: Perspectives on a Modern Catastrophe. Lanham, MD: Rowman and Littlefield.

Fischer, H. W., III. (1998). *Response to disaster: Fact versus fiction and its perpetuation: The sociology of disaster* (2nd ed.). New York: University Press of America.

Fox News. (9/2/2005). Bush to tour Katrina destruction on Friday. Retrieved from http://www.foxnews.com/story/0,2933,168127,00.html. Accessed on 26 Jul 2011.

Gabe, T., Falk, G., & McCarty, M. (2005). *Hurricane Katrina: Social-demographic characteristics of impacted areas*. Washington, DC: Congressional Research Service.

Gray, J., & Wilson, E. (1984). Looting in disasters: A general profile of victimization. Working Paper 71, Colombus, OH: The Ohio State University Disaster Research Center.

Kane, R. (2002). The social ecology of police misconduct. *Criminology, 40*(4), 867–896.

Keizer, G. (3/14/2011). Criminal kick off Japanese disaster scams at record speed. Retrieved from http:/www.computerworld.com/s/article/9214518/Criminals_kick_off_Japanese_disaster_scams_at_record_speed. Accessed on 19 Jan 2012.

Kirk, D. S. (2009). A natural experiment on residential change and recidivism: Lessons from Hurricane Katrina. *American Sociological Review, 74*(3), 484–504.

Lee, T. (8/26/10). Rumor to Facts in Tales of Post-Katrina Violence. *New York Times*, retrieved online from http://www.nytimes.com/2010/08/27/us/27racial.html?pagewanted=all. Accessed on 8 Aug 2011.

Lee, T. (8/27/10). Inquiries give credence to reports of racial violence after Katrina. Retrieved from http://query.nytimes.com/gst/fullpage.html?res=9404E6D6173BF934A1575BC0A9669D8B63&pagewanted=1.Accessed on 27 Jul 2011.

Lindell, M. K., & Prater, C. S. (2003). Assessing community impacts of natural disasters. *Natural Hazards Review, 4*(4), 176–185.

Miller, D. S. (2006). Visualizing the corrosive community: Looting in the aftermath of Hurricane Katrina. *Space and Culture, 9*(1), 71–73.

National Guard Bureau (2006). Hurricane Katrina: national guard's finest hour. Retrieved online from http://www.ng.mil/news/archives/2006/08/082806-Finest_Hour.aspx. Accessed on 8 Aug 2011.

National Oceanic and Atmospheric Administration: National Climatic Data Center. (2011). 1980–2010 billion dollar U.S. weather disasters. http://lwf.ncdc.noaa.gov/img/reports/billion/disasters2010.pdf. Accessed on 27 Jul 2011.

New Orleans Police Department, Administrative and Support Bureau. (2004 & 2006). U.C.R. Reportable Crimes, yearly crime statistics. Retrieved online from http://www.nola.gov/GOVERNMENT/NOPD/Crime-Stats/Crime-Stats/. Accessed on 8 Aug 2011.

Potter, H. (2007). Reframing crime in a disaster: Perception, reality and criminalization of survival tactics among African-Americans in the aftermath of Katrina. In H. Potter (Ed.), *Racing the storm: Racial implications and lessons learned from Hurricane Katrina*. Lanham, MD: Rowman and Littlefield.

Quarantelli, E.L., & Dynes, R.R. (1972). When disaster strikes (it isn't much like what you've heard and read about). *Psychology Today*, 5, 66–70.

Quarantelli, E.L. (1987). Disaster Studies: An analysis of the social historical factors affecting the development of research in the area. *International Journal of Mass Emergencies and Disasters, 5*(3), 285–310.

Quarantelli, E.L. (2006). Looting and other criminal behavior in Hurricane Katrina: Atypical and complex but seen before in other catastrophes. Preliminary paper, University of Delaware, Disaster Research Center, Newark.

Rojek, J., & Smith, M. (2007). Law enforcement lessons learned from Hurricane Katrina. *Review of Policy Research, 24*(6), 589–608.

Roberts, Y., Mitchell, M. J., Witman, M., & Taffaro, C. (2010). Mental health symptoms in youth affected by Hurricane Katrina. *Professional Psychology: Research and Practice, 41*(1), 10–18.

Shaw, C. R., & McKay, H. D. (1942). *Juvenile delinquency in urban areas*. Chicago: University of Chicago Press.

Tierney, K., Bevc, C., & Kuligowski, E. (2006). Metaphors matter: Disaster myths, media frames, and their consequences in Hurricane Katrina. *Annals of the American Academy of Political and Social Science, 604*(1), 57–81.

U.S. Climate Change Science Program. (2008). Weather and climate extremes in a changing climate: Regions of focus: North America, Hawaii, Caribbean, and U.S. Pacific Islands. *U.S. Climate Change Science Program and the Subcommittee on Global Change Research*. Department of Commerce, NOAA's National Climatic Data Center, Washington, DC, USA, 164 pp.

U.S. Department of Justice. (8/28/2007). Fact Sheet: Department of Justice Efforts to Rebuild New Orleans after Hurricane Katrina. Department of Justice. http://www.justice.gov/opa/pr/2007/August/07_opa_666.html. Retrieved on 27 Jul 2011.

U.S. Department of Justice. (2011). United States Department of Justice Civil Rights Division. Investigation of the New Orleans Police Department. http://www.justice.gov/crt/about/spl/nopd_report.pdf. Retrieved on 27 Jul 11.

Varano, S. P., Schafer, J. A., Cancino, J. M., Decker, S. H., & Greene, J. R. (2010). A tale of three cities: Crime and displacement after Hurricane Katrina. *Journal of Criminal Justice, 38*, 42–50.

Weaver, J. (2005). Disconnected families coping with fear, stress. MSNBC. Retrieved online from http://www.msnbc.msn.com/id/9172856/. Accessed on 2 Sep 2005.

Chapter 11
Uncontrollable Nuclear Power Accidents and Fatal Environmental Harm: Why We Have Not Been Ready for the Impacts of Climate Change*

Noriyoshi Takemura

Introduction

In Japan the big earthquake and tsunami on the 11th of March 2011 triggered a major nuclear power plant accident, which has caused and will continue to cause tremendous damage and harm. This "accident", however, is not a natural disaster but a human-made disaster and can be appreciated as one of the most serious crimes committed by the state-corporate complex in Japan's history.

From the point of view of climate change, we can anticipate many more natural disasters such as tsunami and cyclones as the earth warms up and new climatic conditions emerge. In the midst of more frequent and more intense climate related events, it is essential that people everywhere be prepared for the potential advent and devastating aftermath of profound natural disasters of this kind. Yet, as this chapter demonstrates, certain vested interests in the corporate sphere and the state arena are already stifling adequate knowledge, discussion and action around these sorts of questions. Insofar as this remains the case, substantive measures to mitigate and adapt to climate change will accordingly be diminished.

The following problems are discussed in this chapter:

1. Six months after the Fukushima nuclear power plant accident, a new argument about safety and evacuation appeared: some of the exposure could have been prevented if officials had released the computer stored data sooner. In other words, while the radiation spread, the Japanese government kept silent.

---

*This paper is part of "the research on global environmental crises and complexity green criminology", Grant-in-Aid of Scientific Research (c) 2008–2011, and is based on the presentation at the 11th Annual Conference of European Society of Criminology, Vilnius, Lithuania, 21–24 September 2011.

N. Takemura (✉)
Department of Law, Toin University of Yokohama,
1614 Kuroganecho, Aoba-ku, Yokohama, Kanagawa 225-8502, Japan
e-mail: ntakemura@cc.toin.ac.jp

R. White (ed.), Climate Change from a Criminological Perspective, 185
DOI 10.1007/978-1-4614-3640-9_11, © Springer Science+Business Media New York 2012

2. Flawed nuclear power policies and the practices of the industrial-official-academic triad directly contributed to the cause of this catastrophic accident. The political economy of Japanese nuclear power plants is considered through analyzing the historical and social context of their construction and management, as well as through comparing this accident with nuclear power plant accidents in other parts of the world (such as Three Mile Island and Chernobyl).
3. This world-shaking catastrophic disaster has raised serious doubts about the renaissance of nuclear power as a "safe panacea" against global warming. Learning from the Chernobyl disaster and its long-lasting negative effects, what should be done with Fukushima Daiichi nuclear power disaster now and for the future is considered.
4. Leaks from the Fukushima nuclear power plant have caused, are causing, and will cause tremendous radioactive contaminations (of human bodies, foods, air, soil, ocean and so on), and thus contribute to fatal environmental crimes. Its unprecedentedly huge effects are examined.

The chapter concludes by discussing the implications of the harms so described from the point of view of critical green criminology.

While the Radiation Spread, the Japanese Government Kept Silent

We were in the worst place but didn't know it.

Onishi and Fackler (2011a, b) describe the state of straying residents without accurate information right after the accident happened.

The day after a giant tsunami set off the continuing disaster at the Fukushima Daiichi nuclear plant, thousands of residents at the nearby town of Namie gathered to evacuate.

Given no guidance from the Japanese government, town officials led the residents north, believing that winter winds would be blowing south and carrying away radioactive emissions. For three nights, while hydrogen explosions at four of the reactors spewed radiation into the air, they stayed in a district called Tsushima where the children played outside and some parents used water from a mountain stream to prepare rice.

In fact, the winds had been blowing directly toward Tsushima, and town officials would learn 2 months later that a government computer system designed to predict the spread of radioactive releases had been showing just that.

But the forecasts were left unpublicised by bureaucrats in Tokyo, operating in a culture that sought to avoid responsibility and criticism. Japanese political leaders at first did not know about the system and later played down the data, apparently fearful of having to significantly enlarge the evacuation zone, and thereby acknowledge the accident's severity (Onishi and Fackler 2011b).

Computers Tallied Data, But Officials in the Know Withheld the Information

Onishi and Fackler (2011b) insist that there is a strict control of accident information.

Some current and former government officials have admitted that the Japanese authorities engaged in a pattern of withholding damaging information and denying facts of the nuclear disaster in order to limit the size of costly and disruptive evacuations in land-scarce Japan and to avoid public questioning of the politically powerful nuclear industry. As the nuclear plant continues to release radiation, some of which has slipped into the nation's food supply, public anger is growing at what many see as an official campaign to play down the scope of the accident and the potential health risks.

Meltdowns at three of the Fukushima Daiichi six reactors went officially unacknowledged for months. In one of the most damning admissions, nuclear regulators said in early June that inspectors had found tellurium 132, which experts call telltale evidence of reactor meltdowns, a day after the tsunami, but did not tell the public for nearly 3 months. For months after the disaster, the government flip-flopped on the level of radiation permissible on school grounds, causing continuing confusion and anguish about the safety of schoolchildren in Fukushima (Onishi and Fackler 2011b).

Too Late, but Have yet to ...

It is possible to say that a lack of information disclosure has caused a wide range exposure and lots of problems, according to Onishi and Fackler (2011b).

The timing of many admissions, coming around late May and early June, when inspectors from the International Atomic Energy Agency visited Japan and before Japan was scheduled to deliver a report on the accident at an I.A.E.A. conference, suggested to critics that Japan's nuclear establishment was coming clean only because it could no longer hide the scope of the accident. On July 4, the Atomic Energy Society of Japan, a group of nuclear scholars and industry executives, said that it is extremely regrettable that this sort of important information was not released to the public until 3 months after the fact, and only then in materials for a conference overseas.

The group added that the authorities had yet to disclose information like the water level and temperature inside reactor pressure vessels that would yield a fuller picture of the damage. Other experts have said that the government and Tokyo Electric Power have yet to reveal plant data that could shed light on whether the reactor's cooling systems were actually knocked out solely by the 14-m tsunami, as officials have maintained, or whether damage from the earthquake also played a role, a finding that could raise doubts about the safety of other nuclear plants in a nation as seismically active as Japan.

Critics inside and outside the Japanese government argue that some of the exposure could have been prevented if officials had released the data sooner. Mr. Kosako, who studied the Soviet response to the Chernobyl crisis, quickly advised the chief cabinet secretary to use Speedi, which uses measurements of radioactive releases, as well as weather and topographical data, to predict where radioactive materials could travel after being released into the atmosphere. Speedi had been designed in the 1980s to make forecasts of radiation dispersal that were supposed to be made available at least to local officials and rescue workers to guide evacuees away from radioactive plumes. Mr. Baba, the mayor of Namie, said that if the Speedi data had been made available sooner, townspeople would have chosen to flee to safer areas, but we didn't have information.

Information should be disclosed correctly and quickly so that the people can make judgments (Onishi and Fackler 2011b).

Flawed Nuclear Power Policies and Practices

Warning Ignored at Crippled Reactor

Tabuchi et al. (2011) criticize unhealthy ties between plant operators and regulators. Their relation should be not symbiotic but independent.

Just a month before a powerful earthquake and tsunami crippled the plant at the center of Japan's nuclear crisis, government regulators approved a 10-year extension for the oldest of the six reactors at the power station despite warnings about its safety.

The committee reviewing extensions pointed to stress cracks in the backup-powered engines at Reactor No. 1 at the Fukushima Daiichi plant, according to a summary of its deliberations. The cracks made the engines vulnerable to corrosion from seawater and rainwater. The engines are thought to have been knocked out by the tsunami, shutting down the reactor's vital cooling system.

Several weeks after the extension was granted, the company admitted that it had failed to inspect 33 pieces of equipment related to the cooling systems, including water pumps and diesel generators, at the power station's six reactors, according to findings published shortly before the earthquake.

Less than 2 weeks later, the earthquake and tsunami set off the crisis at the power station.

The decision to extend the reactor's life, and the inspection failures at all six reactors, highlight what critics describe as unhealthy ties between power plant operators and the Japanese regulators that oversee them. Expert panels like the one that recommended the extension are drawn mostly from academia to backstop bureaucratic decision-making and rarely challenge the agencies that hire them.

Because public opposition to nuclear power makes it hard to build new power plants, nuclear operators are lobbying to extend their reactors' use beyond the 40-year statutory limit, despite uneven safety records and a history of cover-ups.

The government, eager to expand the use of nuclear energy and reduce the reliance on imported fossil fuels, has been largely sympathetic. Such extensions are also part of a global trend in which aging plants have been granted longer lives (Tabuchi et al. 2011).

Japan's Original Nuclear Problem: A Culture of Collusion

The "Village"

Onishi and Belson (2011) discuss the nature of the "nuclear power village" which makes a point of protecting their interests rather than improving safety.

In Japan, nuclear power's main players are more interested in protecting their interests than improving safety. The opaque network of connections between the nuclear industry and government officials is now popularly referred to as "nuclear power village"—an expression connoting the collusive interests that underlie the nuclear establishment's push to expand the industry, despite the discovery of active fault lines under plants, new projections about the size of tsunami and a long history of cover-ups of safety problems.

As in any Japanese village, the likeminded (nuclear industry officials, bureaucrats, politicians and scientists) have prospered by rewarding one another with construction projects, lucrative positions and political, financial and regulatory support. The few who are openly skeptical of nuclear power's safety become village outcasts, losing out on promotions and backing. Until recently, it had been considered political suicide to even discuss the need to reform an industry that appeared less concerned with safety than maximising profits.

Though it is charged with oversight, the Nuclear and Industrial Safety Agency is part of the Ministry of Trade, Economy and Industry, the bureaucracy charged with promoting the use of nuclear power. Over a long career, official are often transferred repeatedly between oversight and promotion divisions, blurring the lines between supporting and policing the industry.

Influential bureaucrats tend to side with the nuclear industry because of a practice known as "amakudari", or descent from heaven. Widely practiced in Japan's main industries, amakudari allows senior bureaucrats, usually in their 50s, to land cushy jobs at the companies they once oversaw (Onishi and Belson 2011).

Bipartisan Support

Onishi and Belson mention that Japanese nuclear power policy mobilizes almost all the members of nuclear power industries and major political parties.

The political establishment, one of the great beneficiaries of the nuclear power industry, has shown little interest in beefing up safety. In fact, lax oversight serves the establishment's interests. Costly renovations get in the way of building new plants, which create construction projects, jobs and generous subsidies to host communities.

The Liberal Democrats, which governed Japan nearly without interruption from 1955 to 2009, have close ties to the management of companies related to the nuclear industry. The governing Democratic Party has been backed by labor unions, which tend to be close to management. Both parties are captive to the power companies, and they follow what the power companies want to do.

What is more, Japan would make the sale of nuclear reactors and technology the central component of a long-term export strategy to energy-hungry developing nations. A new company, the International Nuclear Energy Development of Japan, was created to do just that. Its shareholders were composed of the country's nine main nuclear operators, three manufactures of nuclear reactors and the government itself.

The nuclear power village was going global with new company. The government took a 10% stake. Tokyo Electric took the biggest, with 20%, and one of its top executives was named the company's first president (Onishi and Belson 2011).

Court Rulings on Nuclear Plants Called into Question

According to Onishi and Fackler (2011a), we should raise a doubt about the attitude of Japanese court that have sided with the authorities and ruled that reactors are safe.

The nuclear power plant could not withstand the kind of major earthquake that new seismic research now suggested was likely, lawyers argued. If such a quake struck, electrical power could fail, along with backup generators, crippling the cooling system, the lawyers predicted. The reactors would then suffer a meltdown and start spewing radiation into the air and sea. Tens of thousands in the area would be forced to flee.

The lawsuits reveal a disturbing pattern in which operators underestimated or hid seismic dangers to avoid costly upgrades and keep operating. And the fact that virtually all these suits were unsuccessful reinforces the widespread belief in Japan that a culture of collusion supporting nuclear power, including the government, nuclear regulators and plant operators, extends to the courts as well.

The crisis at the Fukushima Daiichi nuclear power station has renewed criticism of rulings on the safety of nuclear facilities handed down by the nation's courts. For years, residents in communities near nuclear plants have fought against the state and utilities over the safety of the facilities, contending that the plants could not stand up to major earthquakes. But the courts have almost invariably sided with the authorities, ruling the reactors are safe. The court is partially to blame for giving tacit approval for the operation of the plant (Onishi and Fackler 2011a).

Selling the Nuclear "Safety Myth" in Japan

Onishi analyzes the process and the way how the nuclear "safety myth" has been established and penetrated into Japanese people.

Over several decades, Japan's nuclear establishment has devoted vast resources to persuade the public of the safety and necessity of nuclear power. Plant operators built lavish, fantasy-filled public relations buildings that became tourist attractions. Bureaucrats spun elaborate advertising campaigns through a multitude of organisations established solely to advertise the safety of nuclear plants. Politicians pushed through the adoption of government-mandated school textbooks with friendly views of nuclear power.

The result was the widespread adoption of the belief, called the "safety myth", that Japan's nuclear power plants were absolutely safe. Japan single-mindedly pursued nuclear power even as Western nations distanced themselves from it. The belief helps explain why in the only nation to have been attacked with atomic bombs, the Japanese acceptance of nuclear power was so strong that the accidents at Three Mile Island and Chernobyl barely registered. Even with the nuclear crisis at the Fukushima Daiichi nuclear power plant, the reaction against nuclear power has been much stronger in Europe and the United States than in Japan itself.

Because of this widespread belief in Japanese plant's absolute safety, plant operators and nuclear regulators failed to adopt proper safety measures and advances in technology, like emergency robots, experts and government officials acknowledge.

What became clear was that Japan lacked some of the basic hardware to respond to a nuclear crisis and had to look abroad for help (Onishi 2011).

Rural Japan's Nuclear Trade-Off

Communities Rely on Plants for Funds and Jobs

Fackler and Onishi (2011) mention that rural societies are dependent on nuclear power plants in terms of funds and jobs.

Tokyo has been able to essentially buy the support, or at least the silent acquiescence, of rural communities by showering them with generous subsidies, payouts and jobs. In 2009 alone, Tokyo gave $1.15 billion for public works projects to communities that have electric plants, according to the Ministry of Economy, Trade and Industry. Experts say the majority of that money goes to communities near nuclear plants.

This is just the tip of the iceberg, as the communities also receive a host of subsidies, property and income tax revenues, and compensation to individuals, and even "anonymous" donations to local treasuries that are widely believed to come from plant operators.

Unquestionably, the aid has enriched rural communities that were rapidly losing jobs and people to the cities. But the largess has also made communities depend on central government spending, and thus unwilling to rock the boat by pushing for robust safety measures.

In a process that critics have likened to drug addiction, the flow of easy money and higher-paying jobs quickly replaces the communities' original economic basis, usually farming or fishing.

Experts and residents say this dependency helps explain why, despite the legacy of Hiroshima and Nagasaki, and the accidents at the Three Mile Island and Chernobyl nuclear plants, Japan never faced the levels of popular opposition to nuclear power seen in the United States and Europe. Towns become enmeshed in the same circle that includes politicians, bureaucrats, judges and nuclear industry executives, which has relentlessly promoted the expansion of nuclear power over safety concerns (Fackler and Onishi 2011).

A Flow of Cash and Code of Silence

Fackler and Onishi (2011) also mention that the structure of dependency deprives rural communities and people of the opportunity for criticizing nuclear power plants.

Much of flow of cash was the product of the Three Power Source Development Laws, a sophisticated system of government subsidies created in 1974 by Kakuei Tanaka, the powerful prime minister who shaped the Japanese nuclear power landscape and used big public works projects. The law required all Japanese power consumers to pay, as part of their utility bill, a tax that was funneled to communities with nuclear plants.

Political experts say the subsidies encourage not only acceptance of a plant but also its expansion over time. That is because subsidies are designed to peak soon after a plant or reactor becomes operational and then decline. As the subsidies continue to decline over the lifetime of a reactor, communities come under pressure to accept the construction of new ones. The local community gets used to the spending they got for the first reactor, and the second, third, fourth and fifth reactors help them keep up. If you take them once, you'll definitely want to take them again.

This structure of dependency makes it impossible for communities to speak out against the plants or nuclear power (Fackler and Onishi 2011).

Nuclear Power Plant as "Safe Panacea" or "Catastrophe"

Japan Rates Nuclear Issues at the Level of Chernobyl

Alert Level 7

Mackey (2011) criticizes Japanese officials for playing down the possibility of large amount radiation release.

Japan has raised its assessment of the accident at the crippled Fukushima Daiichi nuclear power plant to the worst rating on an international scale, putting the disaster on par with the 1986 Chernobyl explosion, the Japanese nuclear regulatory agency said on 12 April 2011.

The decision to raise the alert level to 7 from 5 on the scale amounts to an admission that the accident at the nuclear facility is likely to have substantial and long-lasting consequences for health and for the environment. Some in the nuclear industry have been saying for weeks that the accident released large amounts of radiation, but Japanese officials had played down this possibility.

The new estimates by the Japanese authorities suggest that the total amount of radioactive materials released so far is equal to about 10% of that released in the Chernobyl accident, said the deputy director general of Japan's nuclear regulator, the Nuclear and Industrial Safety Agency (Mackey 2011).

Tainted Water Release into Pacific

Belson (2011) warns a danger of contaminated water release into the Pacific Ocean which would result in the intake of radiation through consuming seafood.

Tokyo Electric Power said on 4 April 2011 that it would release almost 11,500 tons of water contaminated with low levels of radiation from the Fukushima Daiichi nuclear plant into the Pacific Ocean, as workers struggled to contain the increasing amounts of dangerous runoff resulting from efforts to cool the plant's damaged reactors.

The water that will be released contains about 100 times the legal limits of radiation, Tokyo Electric Power said. Consuming seafood caught in the area every day for a year would result in the intake of about 0.6 millisieverts of radiation, or about a quarter of the average annual exposure to radiation in Japan, a company spokesman said (Belson 2011).

Nuclear Team Scolds Japan for a Lack of Preparation

Fackler (2011) mentions the lack of preparation against accident and regulatory independence.

Japan underestimated the danger of tsunami and had failed to prepare adequate backup systems at the Fukushima Daiichi nuclear plant, a team of inspectors from an international nuclear regulator said on 1 June 2011, in a critical report.

In a preliminary report on the nuclear crisis that echoed earlier criticisms of Japan's inadequate safety measures, the team from the Vienna-based International Atomic Energy Agency also called for stronger regulatory oversight, saying that steps should be taken to ensure that "regulatory independence and clarity of roles are preserved in all circumstances." This seems to repeat a widely held criticism in Japan that collusive ties between regulators and industry led to weak oversight and a failure to ensure adequate safety levels at the plant (Fackler 2011).

Chernobyl's Lessons for Japan

Garrett (2011) tells us what should be learned from Chernobyl accident.

In 2006, a multi-agency panel of United Nations experts estimated that 200,000 square miles of Eastern Europe were blanketed with fallout, five million residents of the area were exposed, and 100,000 people continue to receive radiation contamination from their food and environment.

Following the Chernobyl meltdown, the nearby town of Pripyat was evacuated, and it remains a ghost town inside a so-called Alienation Zone around the nuclear power plant. Though some Ukrainians continue living in the zone, radiation levels detected in soil samples and local flora and fauna remain high, and food in the form of local animals or mushrooms is still considered dangerous for human consumption.

Nearly 7,000 workers still make their way to the plant every day, maintaining safety operations.

According to the Center for Russian Environmental Policy in Moscow and its 1996 publication, "Consequences of the Chernobyl Catastrophe: Environmental Health," levels of plutonium-238 isotope radiation found years after the Ukranian nuclear accident in the village of Starye Bobrovichi, in the Bryansk Oblast of south-western Russia, were 89 times greater than the combined impact of all 1960 surface nuclear bomb tests (both Soviet and Western).

Local fish and wildlife are heavily contaminated with radioactive cesium-137 and local biologists claim evidence of high levels of mutations in both flora and fauna.

Thirteen years after the Chernobyl disaster, the incidence of pediatric thyroid cancer was 52 times the region's pre-1986 level. In Belarus, where the fallout blew, it was 113 times higher. In the immediate area surrounding Chernobyl, the incidence of pediatric and adult thyroid cancer remains the highest found anywhere in the world, more than 500 times the pre-1986 levels for the region.

Chernobyl showed that the environmental absorption of fallout results in persistent radiation that will affect plants, wildlife, soils and water for thousands of years (Garrett 2011).

To Improve Nuclear Safety for our Future

Garrett (2011) alerts us to catastrophic consequences of nuclear power plant accidents, and insist that safety must be paramount.

Over a quarter of a decade ago, the explosion at Chernobyl cast a radiation cloud over Europe and a shadow around the world. Today, the tragedy at the Fukushima Daiichi nuclear power plant continues to unfold, raising popular fears and difficult questions.

To many, nuclear energy looks to be a clean and logical choice in an era of increasing resource scarcity and due to the pressures to reduce carbon emissions because of global warming. Yet the record requires us to ask: have we correctly calculated its risks and costs? Are we doing all we can to keep the world' people safe. Because the consequences are catastrophic, safety must be paramount. Because the impact is transnational, these issues must be debated globally (Garrett 2011; Shrader-Frechette 2010; Sovacool and Cooper 2008; Stoett 2003; Netzer and Steinhilber 2011; Schneider et al. 2011; IAEA 2011).

The secretary general of the United Nations, Ban Ki-moon, has put forward a five-point strategy to improve nuclear safety for our future:

- First, it is time for a top to bottom review of current safety standards, both at the national and international levels.
- Second, we need to strengthen the work of the International Atomic Energy Agency on nuclear safety.
- Third, we must put a sharper focus on the new nexus between natural disasters and nuclear safety. Climate change means more incidents of freak and increasingly severe weather. With the number of nuclear facilities set to increase substantially over the coming decades, our vulnerability will grow.
- Fourth, we must undertake a new cost-benefit analysis of nuclear energy, factoring in the costs of disaster preparedness and prevention as well as cleanup when things go wrong.
- Fifth, we need to build a stronger connection between nuclear safety and nuclear security. At a time when terrorists seek nuclear materials, we can say that a nuclear plant that is safer for its community is also more secure for the world.

Issues of nuclear power and safety are no longer purely maters of national policy alone. They are a matter of global public interest. We need international standards for construction, agreed guarantees of public safety, full transparency and information-sharing among nations (Ban Ki-Moon 2011).

Tremendous Radioactive Contaminations

Contamination of Air, Water, Soil, Fish, Meat, and ...

Jolly and Grady (2011), Pollack et al. (2011), and Tabuchi et al. (2011) vividly describe the contamination of air, water, soil, fish, meat and so on which appear one after another after the accident.

Tabuchi et al. (2011) insists that questions about radiation tests would have caused rising fears.

Radiation in Tokyo Water Deemed Unsafe for Infants

Radioactive iodine detected in the capital's water supply prompted a warning for infants on 23 March 2011. The head of water purification for the Tokyo water department said at a televised news conference that infants in Tokyo and surrounding areas should not drink tap water. He said that iodine 131 had been detected in water samples at a level of 210 becquerels per liter. The recommended limit for infants is 100 becquerels per liter, or quart (Jolly and Grady 2011).

Radiation Safety Standards for Fish

Japan's government announced its first radiation safety standards for fish on 5 April 2011. The new standards come after fish caught off the coast of north-east Japan, halfway between the Fukushima Daiichi Nuclear Power Station and Tokyo, were found to contain high levels of radioactive iodine 131. The fish were caught on 1 April 2011 and contained 4,080 becquerels per kilogram, or about 2 lb, of iodine 131. The new standard allow up to 2,000 becquerels per kilogram of iodine 131, the same standard used for vegetables.

Tokyo Electric said on 5 April 2011 that it had found radioactive iodine 131 in seawater samples to be at 200,000 becquerels per cubic centimeter, or five million times the legal limit. The samples were collected on 4 April 2011 near the water intake of the No. 2 reactor of the Daiichi plant. The sample also showed levels of cesium 137 to be 1.1 million times the legal limit. Cesium remains in the environment for centuries, losing half its strength every 30 years (Pollack et al. 2011).

Map Shows High Levels of Radiation Near Plant

The first map of ground surface contamination within 80 km of the crippled Fukushima Daiichi nuclear power plant, which was released on 6 May 2011, shows radiation levels higher in some municipalities than those in the mandatory relocation zone around the Chernobyl nuclear plant.

The total are contaminated with radiation from the Fukushima Daiichi nuclear power plant is estimated to be about 800 km^2, or nearly 40% of the size of Tokyo, according to the radiation map created by the Ministry of Education, Culture, Sports, Science and Technology and the US Department of Energy.

The total area is about one-tenth the size of the area contaminated in the Chernobyl nuclear accident.

About 370,000–630,000 terabecquerels of radioactive substances are estimated to have been released by the crippled plant, according to a report by the Nuclear and Industrial Safety Agency and the Nuclear Safety Commission of Japan. That radiation amount equals about one-tenth the amount released from the Chernobyl nuclear power plant, whose nuclear reactor was destroyed.

From 1 April through 6 April, 520 tons of highly radioactive water were discharged from the plant, while radiation totaling 4,700 terabecquerels was leaked, according to Tokyo Electric Power, raising concerns over severe environmental pollution.

As of 13 May 2011 cesium and iodine were still being detected 40 kilometer offshore of the plant. Levels of cesium higher than the safety standard have been found in bamboo shoots, shiitake mushrooms and mountain vegetables (Jolly and Grady 2011).

Tea Shows Excessive Radiation

Tea leaves containing concentrations of radioactive cesium exceeding national safety standards were found in five more refineries, officials of the Shizuoka prefectural government have said. That brings total number of refineries with cesium-laden tea leaves to six.

After the announcement on 14 June 2011, the prefectural government immediately ordered operators of the five plants to suspend shipments and collect those that have been already shipped.

The health ministry's Yokohama Quarantine Station inspected the "seicha", or refined tea leaves, of 20 refineries in the Warashina district of Aoi Ward, in Shizuoka, on 13 and 14 June 2011. It detected concentration levels of radioactive cesium in five plants ranging from 581 to 654 becquerels per kilogram, higher than the national safety standard of 500 becquerels (Tabuchi 2011b).

Tainted Beef, Few Tests and Rising Fears

Japanese agricultural officials say meat from at least 132 cattle that were likely to have been contaminated with radioactive cesium has made its way to supermarkets and restaurants across Japan in the past weeks, according to the IHT on 19 July 2011. Officials say the cattle ate hay that had been stored outside and might have gotten contaminated by radioactive rainwater.

Contaminated hay has been found at farms 140 km from the crippled Fukushima Daiichi plant, raising fears that the radioactive fallout has reached a bigger area than first suspected.

The government's disclosure has also underscored how little Japan has been able to do to control the spread of radioactive material into the nation's food supply. A severe shortage of testing equipment, and local governments that are still swamped with disaster relief, have meant that only a small percentage of farm products grown in the region get checked for radiation (Tabuchi 2011b).

Food Worries Grow Along with Questions About Radiation Tests

Three months after the earthquake and tsunami crippled the plant, Japan does not appear to have a comprehensive food-testing regimen, said Peter Burns, the former chairman of the UN Scientific Committee on the Effects of Atomic Radiation. Prolonged exposure to radiation in the air, ground and food can cause leukemia and other cancers, according to the World Nuclear Association.

Products including spinach, mushrooms, bamboo shoots, tea, milk, plums and fish have been found to be contaminated with cesium and iodine as far as 360 km from the station. Contamination was detected in 347 food samples from eight prefectures by June 9, according to the Ministry of Health, Labor and Welfare.

Over a 3 month period, more than 4,850 samples from 22 prefectures were tested for radiation. The voluntary tests are conducted by prefectural governments in cooperation with local farmers. There is no centralized checking system, and many small farms are not tested.

Restrictions were imposed on food shipments from Fukushima, Shizuoka, Tochigi, Gunma, Ibaraki, Chiba and Kanagawa. Shipments of some green tea were halted in four prefectures (Tabuchi 2011b).

Lower pay, Fewer Benefits, More Radiation

Who wants to get the mission which is attended by the risk of radiation contamination? Tabuchi (2011b) tries to make clear what kind of people nuclear workers become and what a magnet for them is.

Day Workers Must Tackle Plant's Riskiest Tasks

The nuclear workforce is emblematic of Japan's two-tiered work force, with an elite class of highly paid employees at top companies and a subclass of laborers who work for less pay, have less job security and receive fewer benefits. Such labor practice have both endangered the health of these workers and undermined safety at Japan's 55 nuclear reactors.

In Japan's nuclear industry, the elite are operators like Tokyo Electric and the manufacturers that build and help maintain the plants like Toshiba and Hitachi. But under those companies are contractors, subcontractors and sub-subcontractors with wages, benefits and protection against radiation dwindling with each step down the ladder. Some workers are hired from construction sites, some are local farmers looking for extra income, and others are hired by local gangsters (Tabuchi 2011a).

Crippled Plant a Magnet for Job Seekers

The company refused to say how many contract workers have been exposed to radiation. Day laborers are being lured back to the plant by wages that have increased along with the risks of working there. Despite the still-present dangers at Fukushima Daiichi, laborers from across Japan are traveling to the plant in search of work amid Japan's harsh economic downturn.

Tokyo Electric transfers risk to contractors, subcontractors, or sub-sub contractors and their poorly paid, poorly trained employees, endangering their health and undermining safety at Japan's 55 nuclear reactors. There is no work at a nuclear power plant that doesn't involve radiation risks.

Amid the ongoing confusion at the Fukushima Daiichi plant, there has been minimal monitoring of radiation exposure. The Radiation Effect Association, a government-affiliated body tasked with keeping track of radiation exposure levels among Japanese nuclear workers, say that it has not been able to fully track radiation exposure among plant workers past March 11 (Tabuchi 2011a).

Fukushima Radiation Effects Research

Wald and Grady (2011) insist that we need long-lasting researches and studies of radiation effects to people who have contaminated from the nuclear plant accident and will continue to live near the nuclear plant.

Unknowns Plague Japan Radiation Studies

As officials in Japan agonise over what constitutes a safe radiation dose for people who live near the Fukushima Daiichi nuclear reactors, the state of the science has been a daunting problem. Studies on the effects of exposure are based mostly on large doses delivered quickly by atomic bombs, while radiation from the Fukushima disaster is more likely to take the form of small doses delivered over many years.

The general assumption is that when people are exposed to small doses for decades, the incidence of cancer will rise over time. But that prediction is based on extrapolating from data on people who were exposed to acute brief doses when atomic bombs were dropped on Nagasaki and Hiroshima in 1945, not on observing individuals exposed to small doses over decades.

Some researchers argue that all humans are regularly exposed to a low natural level of radiation and that it is not harmful when below a certain threshold, although foetuses may be an exception. Another vocal minority argues that there is statistical evidence for higher cancer rates among people exposed to tiny incremental doses.

Some of the radiation to which people are being exposed around Fukushima is inside the body. It comes from radicactive materials that contaminated their food or water (Wald 2011).

Health Effects of Disaster to be Studied

The Fukushima prefecture government plans to monitor the health effects of radiation leaking from the Fukushima Daiichi nuclear power plant in a program that could follow some residents for as long as 30 years. The study will be unprecedented in its length and in the number of people covered.

Because thorough radiation monitoring of all Fukushima residents would be impossible, only those found to have comparatively high estimated radiation exposure levels will undergo monitoring that includes blood and urine tests.

Despite the scale of the proposed study, it may not be easy to determine clearly what health effects may result from radiation from the Fukushima plant. It has generally been thought that health effects will arise only if an individual is exposed to more than 100 millisievelts of radiation. The general opinion of experts is that the average Fukushima resident was likely to have been exposed to, at most, several dozen millisievelts of radiation. Although that is much higher than the 2.4 millisievelts per year of natural radiation in the air, it is not a level for which health effects have been confirmed in past studies.

The Fukushima study will go beyond any that have been conducted on survivors in Hiroshima and Nagasaki because it will include measurement of internal contamination levels (Grady 2011)

Transgressive Environmental Crime

State-Corporate Crime

The concept "state-corporate crime" was coined by Michalowski and Kramer (2006).

In so far as their function is to support government sponsored atomic power programs, generate export revenue and more generally contribute to economic growth, electric power companies and the atomic industry can be said to have a symbiotic "insider" relationship to the state. Relationships between the state and electric power companies are complex, and can be characterized in terms of mutual interdependence. State-corporate crime is not simply a result of the success or lack of success of the state's control or regulation, but is a product of the symbiotic relationship between state and companies. The prospects for controlling state-corporate crime rest upon the extent to which we can break this symbiosis (Michalowski and Kramer 2006; Potter 2001).

In thinking about state-corporate crime and its control, what is really at issue is not crime; rather, it is an issue of state, corporate, and class power. To turn our attention to the contours and dynamics of class power, and its connection to atomic policies, seems more pressing, since rural communities who have atomic power plants are facing dangers to have radioactive contamination, and job seekers are coming to Fukushima to get a dangerous job. Gaps between rural and urban communities, and job seekers and steady workers, are more and more widening. Thus "the rich get safer and the poor get danger" has become an appropriate aphorism for our times.

Green Criminology and Social Justice

White (2008, 2011) mentions the importance of political economy and historical social context of environmental crime and ecological injustice.

Most of what is happening is directly attributable to negative human intervention and the systemic imperatives, and unnatural consequences, of the global capitalist political economy. Harm is being perpetrated across the earth, although the intensity and form it takes varies depending upon specific region and specific population. Sometimes these harms are acknowledged in the law as criminal offences, but in many cases, they are not. They are simply part of the normal way of doing things, an attitude and a practice that will only lead to disaster for all.

Environmental crime is a topic of growing international interest and importance. The study of environmental crime or harm has expanded considerably in the last decade or so.

Conceptualisation of environmental crime challenges existing paradigms, and confirms the importance of dealing with environmental harms and victimisation in a systematic and theoretically informed manner. Environmental crime is defined variably by different authors, but in ways that include reference to transgressions against humans, environments, and nonhuman animals (White 2008, 2011).

The dynamics and dimensions of environmental crime turn to specific instances of harm and criminal activity. Specific types of environmental harm cannot be separated from the fact that the harm is also specific to very particular groups of people. In other words, environmental harm frequently involves the simultaneous exploitation of particular bio-spheres, of particular plants and animals, and of the poorest, most vulnerable sections of the human community. Environmental crime and ecological injustice take place in within certain historical and social context. One consequence of this is phenomenon of differential victimisation, where different species, bio-spheres and human population groups are disproportionately subject to criminal and/or harmful activities (White 2011; Takemura 2009b, 2010).

Environmental Human Rights and Intergenerational Justice

According to Hiskes (2008), not only our generation but also future generations, all generations have the environmental human rights, and we need a viewpoint of intergenerational justice.

As a matter of justice, all citizens present and future possess the environmental human rights to clean air, water and soil. Future citizens rely on the environmental rights of present citizens being protected for the possibility that theirs will be also protected. The environmental rights of the present depend on the protection of the rights of future citizens. This reciprocity between present and future makes environmental justice across generations possible. Both concepts of intergenerational justice and human rights are newcomers to the politics of environmental preservation. In the context of the increasing likelihood of natural disasters and climate-related incidents, a precautionary approach based upon such concepts is needed now more than ever (Hiskes 2008).

Conclusion

The "accident" at the Fukushima Daiichi nuclear power plant has been menacing not only Japanese people but also people around the world with radioactive contamination. Everyone near the nuclear power plant was evacuated, and deprived of almost of all things they had before the disaster: jobs, homes and normal family life. Its impacts will be felt for many more years to come.

This disaster was structurally caused by symbiotic relations between the government and the Tokyo electric power company who put their interests before the safety of people. This disaster has been causing and will continue to cause unprecedented tremendous radioactive contaminations, and has and will have a great negative effect not only our generation but also future generations for an extremely long time. For example, there is a terrible expectation that diseases of thyroid gland and leukemia will gradually increase in the near future (Takemura 2008, 2009a).

In this sense, it can be evaluated as a human-made great disaster, one of the serious crimes of the state-corporate complex, and a transgressive environmental crime which transgresses time (crossing past, present and future) and space (crossing local, global and cosmic). It is an evil that did not have to happen.

References

Ban, K.-M. (2011). A visit to Chernobyl. *International Herald Tribune*, April 25.

Belson, K. (2011). Filtering of tainted water begins at Japanese plant. *International Herald Tribune*, June 17.

Fackler, M. (2011). Report finds Japan underestimated tsunami danger. *International Herald Tribune*, June 1.

Fackler, M., & Onishi, N. (2011). Japan's nuclear dependency. *International Herald Tribune*, May 31.

Garett, L. (2011). Chernobyl's lessons for Japan. *International Herald Tribune*, March 17.

Grady, D. (2011). Radiation is everywhere, but how to rate harm? *International Herald Tribune*, April 4.

Hiskes, R. P. (2008). *The human right to a green future: Environmental rights and intergenerational justice*. London: Cambridge University Press.

IAEA: International Atomic Energy Agency, Planning and Economic Studies Section, Department of Nuclear Energy. (2011). *Climate change and nuclear power 2011*. Vienna: IAEA.

Jolly, D., & Grady, D. (2011). Anxiety up as Tokyo issues warning on its tap water. *International Herald Tribune*, March 23.

Mackey, R. (2011). A look at the nuclear accident scale. *International Herald Tribune*, April 12.

Michalowski, R. J., & Kramer, R. C. (2006). *State-Corporate crime: Wrongdoing at the intersection of business and government*. Piscataway, NJ: Rutgers University Press.

Netzer, N., & Steinhilber, J. (Eds.). (2011). *The end of nuclear energy? International perspectives after Fukushima*. Germany: Friedrich Ebert Stiftung.

Onishi, N. (2011). 'Safety Myth' left Japan ripe for nuclear crisis. *International Herald Tribune*, June 24.

Onishi, N., & Belson, K. (2011). Culture of complicity tied to stricken nuclear plant. *International Herald Tribune*, April 26.

Onishi, N., & Fackler, M. (2011a). Japanese officials ignored or concealed dangers. *International Herald Tribune*, May 16.

Onishi, N., & Fackler, M. (2011b). Japan held nuclear data, leaving evacuees in Peril. *International Herald Tribune*, August 9.

Pollack, A., Belson, K., & Drew, K. (2011). Company says radioactive water leak at Japan plant is plagged. *International Herald Tribune*, April 5.

Potter, G. W. (Ed.). (2001). *Controversies in white collar crime*. Cincinnati, OH: Anderson.

Schneider, M., Froggatt, A., & Thomas, S. (2011). *The World Nuclear Industry Status Report 2010–2011: Nuclear power in a post-Fukushima World: 25 years after the Chernobyl accident*. Washington, DC: Worldwatch Institute.

Shrader-Frechette, K. (2010). Environmental injustice, climate change, and nuclear power: Flawed standards for ionizing radiation. *Journal of Applied Ethics and Philosophy, 2*, 1–11.

Sovacool, B. K., & Cooper, Ch. (2008). Nuclear nonsense: Why nuclear power is no answer to climate change and the world's post-Kyoto energy challenges. *Wm. And Mary Environmental Law and Policy Review, 33*, 1–119.

Stoett, P. (2003). Toward renewed legitimacy? Nuclear power, global warming, and security. *Global Environmental Politics, 3*, 99–116.

Tabuchi, H. (2011a). Braving heat and radiation for temp job. *International Herald Tribune*, April 10.

Tabuchi, H. (2011b). Radiation-tainted beef spreads through Japan markets. *International Herald Tribune*, July 18.

Tabuchi, H., Onishi, N., & Belson, K. (2011). Japan extended reactor's life, despite warning. *International Herald Tribune*, March 21.

Takemura, N. (2008). Environmental risks/crimes of nuclear power plant and complexity green criminology: Anatomy of problematique around nuclear power plant, earthquake and environment. *Research Bulletin (Toin University of Yokohama), 19*, 81–88.

Takemura, N. (2009a). Criticality of complex system: Nuclear power plant, earthquake and environment – Are nuclear power plants on earthquake-prone islands 'safe panacea' or 'catastrophe'?–. In Y. Shimomura, T. Morishita, & T. Satoh (Eds.), *Memorial Festschrift for Professor Doctor Kuniyuki Yagi: New development of modern criminal law*. Tokyo: Sakai Shoten.

Takemura, N. (2009b). Factitious Catastrophe, global warning, and chaos/complexity green criminology/justice – Tug-of-war: environmental 'injustice' vs. 'green social justice' 1 –. *Research Bulletin (Toin University of Yokohama), 21*, 83–93.

Takemura, N. (2010). 'Criticality of global environmental crises' and prospects of 'chaos/complexity green criminology': Spreading environmental 'injustice' and struggle for 'chaos/complexity green justice'. In H. Kawabata, T. Shiibashi, & K. Kai (Eds.), *Festschrift for Professor Niroku Tateishi's Seventieth Birthday*. Tokyo: Seibundoh.

Wald, M. L. (2011). Radiation's Unknowns Weigh on Japan. *International Herald Tribune*, June 6.

White, R. (2008). *Crimes against nature: Environmental criminology and ecological justice*. Devon: Willan Publishing.

White, R. (2011). *Transnational environmental crime: Toward an eco-global criminology*. London/New York: Routledge.

Chapter 12
A Proposal for a New Vehicle-Based Carbon Tax (V-CART): Vehicle-Based Global Warming Policy and Green Criminology

Michael J. Lynch and Paul Stretesky

Introduction

Policy concerns surrounding global warming are not widely studied in criminology despite the widespread attention focused relationship between climate change and global ecology. We believe that green criminology offers a useful perspective for examining global warming policy issues (Lynch 2007; Lynch and Stretesky 2010; White 2010a). This is the case because green criminology often deals with regulatory forms of social control designed to address environmental pollution (Burns et al. 2008). The purpose of this chapter is to extend the study of green criminology to social control policies thought to slow current upward trends in global warming gases. We do this by examining the existing penalty structure for automobile emissions violations in the United States and explore the potential for a new supplemental carbon tax.

We begin our analysis by briefly examining the relevance of global warming policy for green criminology. We then look at the current corporate average fuel economy (CAFE) policy used to regulate carbon dioxide gases that are released from automobiles that burn fossil fuel. As noted, carbon dioxide is a greenhouse gas that is thought to contribute to global warming. After explaining the limitations of CAFE, we propose an alternative vehicle-based carbon tax (V-CART) that serves as a new social control policy for reducing carbon emissions. We demonstrate how that policy may work and conclude by examining the benefits and costs of such an approach.

M.J. Lynch
Department of Criminology, SOC107, University of South Florida,
Tampa, FL 33620-8100, USA
e-mail: radcrim@tampabay.rr.com

P. Stretesky (✉)
School of Public Affairs, University of Colorado - Denver, Campus Box 142,
1380 Lawrence Street, Suite 500, Denver, CO 80217-3364, USA
e-mail: paul.stretesky@cudenver.edu

R. White (ed.), Climate Change from a Criminological Perspective,
DOI 10.1007/978-1-4614-3640-9_12, © Springer Science+Business Media New York 2012

Green Criminology and the Study of Vehicle Emissions Policy

As this book illustrates, green criminology has begun to address global warming (Lynch 2007; Lynch et al. 2010; Lynch and Stretesky 2010; Stretesky and Lynch 2009; White 2010b). Specifically, in the United States climate change policies are now being addressed by legal mandates. In 2007 (Massachusetts vs. EPA, 549 US 494) the US Supreme Court ruled that the US Environmental Protection Agency had a duty to regulate greenhouse gas emissions under the federal Clean Air Act. This ruling effectively promulgated a legal standard that justifies and legitimizes a green criminological approach (see Chap. 6). To be sure, the basis of green criminology extends beyond legal boundaries. Nevertheless, the US Supreme Court decision has significant implications for the standing of green criminology as a core area of criminological research, and one that can be illustrated to "fit in" with "normalized" forms of criminology that deal with harms that violate law. From a green criminological perspective, studying the effects of global warming and policies related to greenhouse gas emissions can be connected to a broad range of green approaches. There are, for example, global issues associated with greenhouse gas emissions consistent with White's (2010b, 2011) vision of an eco-global green criminology. Likewise, climate change has relevance to Beirne's (1999) non-speciesist approach to harm. And while global warming has cross-species effects, one cannot deny the relevance of an anthropocentric view of global warming since global warming has anthropogenic causes and only the human species can create policies to address this phenomenon. Finally, it is worth repeating White's (2010a, p. xvi) observation that green "criminologists need to ***shape*** and *respond* to the world political agenda" (emphasis added). To date, global warming policies have had much more currency internationally and outside of the US (see Chap. 6). That fact, however, does not excuse green criminology from its failure to address the international scope of harm US global warming policy (or the lack thereof) causes.

We draw upon the legal standard and the notion of harm to specifically focus on vehicle climate change policy in the United States. US consumers comprise less than 5% of the world's population but consume nearly one-quarter of the annual world production of oil, nearly two-thirds of that amount in its vehicle fleet, and produce approximately 20% of total worldwide global warming gases. Thus, a worldwide solution to the problem of global warming requires attention to US efforts to reduce its carbon footprint, and especially to efforts to address vehicle-based global warming gas production (Environmental Protection Agency 2010a). Moreover, since 1990, vehicle-related global warming gas production has been recognized as the fastest growing segment of US global warming (Environmental Protection Agency 2010b).

Existing Policy

In 1975, the US imposed fuel economy regulations known as the Corporate Average Fuel Economy (CAFE) program in response to the oil embargo and energy crisis of 1973–1974. Congress revised the Motor Vehicle Information and Cost Saving Act

(MVICSA) through the Energy Policy Conservation Act (EPCA; (Pub.L. 95-619, 92 Stat. 3206)). Title V of the EPCA established miles per gallon requirements for automobiles and light trucks (including SUVs) which are assessed and regulated relative to a manufacturer's vehicle fleet (for extended discussion see, Burns and Lynch 2002, 2004; for further information see the Code of Federal Regulations, Title 49, Subtitle VI, Chap. 329). Manufacturers are fined $5.50 for every one-tenth of a mile their vehicle fleet falls below a given year's CAFE standard, adjusted for the aggregate mile per gallon violation for the fleet (National Highway Transportation Safety Administration 2011a). The National Highway Traffic Administration (2011b) estimates that more than $785 million dollars in civil penalties have been collected under this program since 1983.

CAFE regulations were originally proposed to address fuel economy standards for automobiles and light trucks in an effort to reduce US consumption of and dependence on foreign oil, and to respond to declining production of domestic oil and depletion of US oil deposits. In recent years under the Obama Administration, an effort has been made to revise CAFE regulations to respond to greenhouse gas emissions, and to modify CAFE regulations for this purpose. These revisions include future substitution of greenhouse gas emission standards for miles per gallon standards and the tightening of emission standards (Environmental Protection Agency 2010c).

As an illustration of the logic behind CAFE as a climate change policy we calculate carbon pollution produced by a vehicle. To do this, it is necessary to employ the chemical equation for octane (C_8H_{18}) and to have knowledge of the combustion process ($[2]C_8H_{18} + [25]O_2$ $[16]CO_2 + [18]H_2O$. see, Zakrzewski 2002) in order to derive the carbon outcome that burning one gallon of gasoline produces approximately 19.6 lb of carbon dioxide. Thus, it is possible to calculate vehicle carbon pollution from this outcome (e.g., a vehicle that travels 12,000 miles and averages 25 mpg uses 480 gallons of gasoline, and hence produces 9,408 lb of carbon dioxide; $(12{,}000/25) \times 19.6 = 9{,}408$). Clearly, reducing the volume of gas a vehicle uses by requiring enhanced fuel efficiency lowers carbon dioxide production.

Since its adoption, CAFE regulations have been widely discussed both in terms of their benefits and limitations (Goldberg 1998; Greene 1997; National Academy of Science 2003, p. 1). We must recognize at the outset of our discussion that CAFE debates, whether pro or con, still serve to legitimate individual transport over public transit. Nevertheless, research suggests that current CAFE policy has a small to modest effect on US-based carbon dioxide pollution by reducing potential carbon dioxide production by imposing CAFE-derived vehicle efficiency standards that have reduced *projected* or future fuel consumption (i.e., current gasoline consumption compared to expected gasoline consumption before the standards went into effect or were raised). For example, in its CAFE impact assessment for future model years (2011–2015), the US National Highway Traffic Safety Administration (2008) indicated that improved CAFE standards could save 50.61 billion gallons of gasoline between 2020 and 2060 *if* all vehicles operating during this period included all known fuel savings technology. Averaged over this 40-year period, the annual savings (1.265 million gallons) is less than 1/10th of 1% of projected annual gasoline consumption.

In addition, the National Academy of Sciences (NAS 2003, p. 3) noted that CAFE regulations have "clearly contributed to increased fuel economy of the nation's light-duty vehicle fleet…" NAS also noted, however, that CAFE regulations can play a greater role in reducing the production of greenhouse gases, particularly carbon dioxide, by enhancing passenger vehicle fuel economy. Whether or not this is true remains a question open to further examination. As NAS pointed out, the extensive volume of global warming gas produced by the US passenger fleet vehicle remains a concern, and it is possible for CAFE to play an enhanced role in this regard. As noted, it has been estimated that approximately 20% of US global warming emissions are vehicle related, though the Environmental Defense Fund (2007) places this figure at 45% based on an estimated release of 314 million metric tons of carbon equivalents in 2004 (DeCicco and Fung 2006).

CAFE Limitations

Prior to examining the V-CART policy, we explore four questions and answers about CAFE. These questions are meant to illustrate the disadvantages of CAFE and the need for an alternative policy which we explore below. First, is CAFE effective at reducing carbon dioxide vehicle emissions? Second, does CAFE provide any educative value for consumers? Third, does CAFE address carbon emissions directly? Fourth, does the CAFE penalties structure make sense?

Is CAFE effective at reducing carbon dioxide vehicle emissions? A substantial body of research examines the effectiveness of CAFE regulations and their impacts on concerns such as driving behavior and vehicle safety. CAFE's critics (Crandall 1985, 1990; Crandall and Graham 1989; Crandall et al. 1986) suggest CAFE has unintended consequences such as increasing the use of motor vehicles as they become more efficient and shifting consumers' preferences to used cars to obtain larger or less expensive vehicles. Goldberg (1998) found that CAFE had a bifurcated price effect (a rise in the price of large vehicles and a decline in price for smaller, more efficient vehicles) and no effect on demand for larger used cars. Research has also examined the effect of CAFE on lowering gasoline prices through increased fuel efficiency (Kahn 1996). Studies drawing on data from the late 1970s and early 1980 provided some evidence of this effect, while studies with more recent data do not (Greene 1997). Later studies, however, found that CAFE did not drive fuel prices, and therefore did not support the assumption of lower fuel prices promoting increased fuel consumption.

Vehicle safety, which has been linked to size and weight (NHTSA 1997) has also been a focus of empirical studies of the impact of CAFE. Some studies find that CAFE lead to vehicle size and weight reductions and reduced safety for cars (Crandall and Graham, 1989) and light trucks (Heavenrich and Hellman, 1996). Others, however, note that vehicle weights have increased since 1983, calling these conclusions into question (Greene 1997, p. 20). Moreover, Greene suggests that it is a vehicle's weight distribution rather than its total weight which ought to be considered in safety assessments, and that consumers who choose larger automobiles may be enhancing their safety while posing greater risks to other drivers. In addition, weight-reduction effects vary by vehicle type. For example, a 100 lb weight reduction for passenger cars increases the risk of fatalities by 1%, while a similar reduction in weight to light trucks reduces fatalities by 0.3% (NHTSA 1997). Others note that weight-reduction effects have been offset by safety innovations (front and side air bags, side-impact beams, and rollover safety standards; Brumbelow et al. 2008; Insurance Institute for Highway Safety 2003).

Does CAFE provide any educative value for consumers? Manufacturers are fined when their fleet of vehicles fails to meet the CAFE standard. Consumers, however, are likely unaware that they indirectly pay a fine when purchasing a vehicle from a manufacturer whose fleet violates CAFE standards. In order to pay the fines and maintain their profit margins, the manufacturer passes along the fine in the price of the vehicle. Since CAFE fines are relatively small compared to the purchase price of a vehicle, these fines may have little effect on consumer behavior. As an example, consider the difference in the final price of two vehicles costing $40,000, one from a manufacturer's fleet that meets CAFE requirements and therefore does not include a fine, and one that under-performs relative to the CAFE standard by 5 miles per gallon. The CAFE fine associated with the second vehicle is $275 (5 mpg × 10 × $5.50) or 0.6875% of the vehicle's price—a cost that is hardly likely to impact the decision to buy a vehicle of this price. Extant research suggests that direct or apparent costs, such as the price of gasoline, have a stronger effect on consumer vehicle choice than indirect CAFE fines (Gramlich 2009; Greene 1997; Van Biesebroeck and Leuven 2010). In addition, research suggests that environmentally conscious consumers are more likely to be influenced by vehicle fuel efficiency and efficiency standards (Van Biesebroeck and Leuven 2010), suggesting that direct costs and knowledge impact consumers. The proposed V-CART model (explained below) takes these elements into consideration while current CAFE regulations have failed to address these concerns.

Does CAFE address carbon emissions directly? CAFE only indirectly addresses global warming in relation to new and future vehicle carbon dioxide reductions gained through enhanced vehicle efficiency standards. In other words, CAFE is not designed to directly address global warming pollution and does so only as a secondary consequence of regulating vehicle efficiency. Because the primary goal of CAFE was to increase fuel efficiency to reduce US dependency on foreign oil, carbon dioxide outputs are only indirectly regulated through CAFE's fuel efficiency requirements. As an example consider the impact of CAFE's indirect carbon pollution reductions accomplished through regulating fuel economy versus the direct tailpipe emission standards found in the Clean Air Act. Direct regulation of tailpipe emissions under the Clean Air Act produced a 60% reduction in harmful emissions since 1970 despite growth in the number of vehicles on the road (DeCicco and Fung 2006, p. 16). While the Clean Air Act helped reduce harmful emissions (carbon dioxide is not included in these standards) vehicle-related carbon dioxide emissions increased by 70% during this same time period (DeCicco and Fung 2006, p. 17).

In short, CAFE policy isolates petroleum-based vehicles as a problem and consequently does not situate the policy response by focusing directly on vehicle-related carbon pollution.

The data and observations about the indirect nature of CAFE regulation on carbon dioxide emissions lead us to the conclusion that CAFE fuel standards provide an ineffective solution to the problem of CO2 emissions.

Does the CAFE penalties structure make sense? In order to answer this question we begin with an example of how CAFE fines apply across two companies and note that, within a company, the policy works much like carbon emission trading schemes. Company MEET sells two models; 50,000 units of Model A with an mpg rating of 20, and 100,000 units of Model B with a 33 mpg rating. One-third of this fleet fails to meet the CAFE requirement for that year (27.5) while two-thirds of the fleet exceeds the requirement. The sale of these two models can be used to offset the deficit in Model A. Model B sales produce a mpg credit of 5.5 mpg per car, or plus 550,000 mpg. Model A produces a deficit of 7.5 mpg per vehicle, or a 375,000 mpg deficit. This deficit is exceeded by Model B's performance, and Company MEET not only satisfies CAFE but also exceeds CAFE, and the company can bank 175,000 mpg for later use. The problem here is that environmentally conscious consumers or consumers who can afford only to purchase the smaller, more efficient vehicle (B) make up for the mileage deficit created by consumers who purchase the inefficient vehicle (A). Thus, so long as there is market inequality or a differential distribution of environmentally conscious buyers, CAFE will fail to penalize consumers who choose inefficient vehicles so long as they buy them from a manufacturer with a diverse product base that takes advantage of this market condition to produce a fleet of vehicles that meets the federal standard.

Consider, however, the consumer who purchases a vehicle from Company FAIL, which sells two vehicles (Model C and D) in equal number with the same mpg as MEET's models, but produces an mpg deficit of 2.2 mpg for 50,000 vehicles once the mileage results for Models C and D are assessed. Under CAFE, Company FAIL is assessed a $6,050,000 penalty, or $121 for each unit of Model C sold. While Model A and C have the same mpg rating, consumers who buy Model C are penalized because of the composition of the fleet sold by Company FAIL. So, despite the fact that these two groups of consumers buy cars with the same environmentally negative effects, one group is penalized and the other is not. Examined in this way, the penalty outcome can be assessed as unfair to consumers who buy an equivalent vehicle from a noncomplying company, and to manufacturer FAIL who, after all produces vehicles with identical mileage efficiency compared to manufacturer MEET. Not only is the outcome unfair, but also some consumers are not held responsible for the negative effect of their choice on the environment, and because the penalty is assessed indirectly as noted earlier, there is little likelihood that they understand that their choice has a financial consequence.

An Alternative Solution

The current CAFE program targets manufacturers and vehicle efficiency technology as solutions to global warming gases by imposing standards that, in theory, ought to generate innovations that improve vehicle efficiency. It is safe to say, however, that CAFE does not address the volume of carbon dioxide that is produced. CAFE also does not allow for emissions reductions that are below technologically achievable levels. Thus, this approach ignores a variety of factors in the chain of events that contribute to global warming which can also be mitigated by a more broadly conceived carbon control policy that is not simply focused on fuel efficiency.

CAFE began as a mechanism for regulating vehicle fuel efficiency to decrease US dependence on foreign oil. Since its inception in the late 1970s, the goals of CAFE have been expanded to recognize environmental pollution concerns, and CAFE is now viewed as a mechanism for reducing US carbon dioxide emissions related to the large fleet of vehicles Americans own and operate. In effect, there is a belief that CAFE can serve purposes other than those it was directly designed to impact. Based on the criticisms reviewed above, we suggest that in order to address carbon dioxide pollution more directly and efficiently, there needs to be an alternative to CAFE that focuses more specifically on carbon dioxide emissions control, and in ways that impact a much broader cross-section of vehicles.

In addition, as we illustrate above, there are a variety of criticisms of the CAFE approach. These criticisms relate to the effectiveness of CAFE as a carbon reduction plan, but also target vehicle use and safety concerns. As shall become evident, our proposal does not suffer from these limitations since it creates a tax that pays the cost of carbon pollution that is equal to a vehicles carbon impact. The fine collected is employed to neutralize carbon outputs, and does not require making vehicles more efficient. Thus, our proposal separates the efficiency and carbon issues and views each as a separate, rather than a combined concern.

In order to illustrate how our proposal works, we begin by reviewing data on the vehicle fleet sold in the US during 2008. Table 12.1 displays CAFE data for the 2008 model year across vehicle manufacturers who sold cars and light trucks in the United States. That table includes National Highway Transportation and Safety Administration and the Environmental Protection Agency data for 20 manufacturers. It includes the number of vehicles each manufacturer sold, the Environmental Protection Agency-derived miles per gallon for each manufacturer's fleet, and our estimate of the fleet tons of carbon dioxide produced by each manufacturer based on the Environmental Protection Agency mileage estimate and average vehicle use (estimated as 12,500 miles per year per vehicle, similar to assumptions made by the Environmental Protection Agency and National Highway Transportation and Safety Administration).

The 20 manufacturers listed in Table 12.1 produced 14,006,358 vehicles. Of those vehicles, however, only 624,827 or 4.46% would be penalized under CAFE requirements (i.e., the manufacturers whose fleets failed to meet the

Table 12.1 Manufacturer, NHTSA division, number of vehicles produced, EPA estimated fleet miles per gallon, and estimated fleet tons of CO_2, 2008 Model Year Vehicles

Manufacturer	Division[a]	*N* Vehicles	Fleet mpg[b]	Fleet tons Carbon dioxide[c]
BMW	(I)	290,060	27.4	1,296,801.095
BMW	(LT)	60,753	22.9	324,988.756
Chrysler	(D)	523,332	29.5	2,173,158.305
Chrysler	(LT)	1,150,183	23.6	5,970,229.555
Chrysler	(I)	1,648	27.4	7,367.883
Daimler	(LT)	88,412	20.6	525,570.971
Daimler	(I)	203,251	26.4	934,115.436
Ferrari	(I)	1,645	16.3	12,362.730
Ford	(D)	700,237	29.8	2,878,491.023
Ford	(LT)	1,265,778	23.6	6,570,245.975
Ford	(I)	231,964	30.6	928,614.052
GM	(D)	1,266,208	29.6	5,240,218.919
GM	(LT)	1,806,528	22.8	9,706,126.316
GM	(I)	128,936	31.4	503,014.650
Honda	(D)	378,575	35.3	1,313,751.771
Honda	(LT)	637,582	25.5	3,062,893.922
Honda	(I)	505,719	33.2	1,865,981.250
Hyundai	(I)	268,293	33.8	972,363.683
Hyundai	(LT)	125,543	25.5	603,098.726
Kia	(I)	171,517	33.4	609,066.841
Kia	(LT)	129,670	24.2	656,387.397
Lotus	(I)	450	29.8	1,849.832
Maserati	(I)	3,884	18.2	26,142.308
Mitsubishi	(LT)	60,458	25.9	285,950.000
Mitsubishi	(I)	140,555	29.6	581,688.769
Nissan	(D)	399,467	33.7	1,452,068.472
Nissan	(LT)	295,768	24.0	1,509,649.167
Nissan	(I)	165,802	29.4	690,841.667
Porsche	(LT)	19,401	19.7	129,640.736
Porsche	(I)	18,124	26.4	84,098.106
Spyker	(I)	50	19.5	314.103
Subaru	(D)	21,159	28.9	89,687.803
Subaru	(LT)	86,342	27.1	390,291.328
Subaru	(I)	86,490	28.5	371,755.263
Suzuki	(LT)	42,559	23.8	219,053.676
Suzuki	(I)	68,676	31.4	267,923.885
Tesla	(D)	800	244.0	401.939
Toyota	(D)	492,973	33.9	1,781,392.109
Toyota	(LT)	1,150,342	23.7	4,945,860.549

(continued)

Table 12.1 (continued)

Manufacturer	Division[a]	*N* Vehicles	Fleet mpg[b]	Fleet tons Carbon dioxide[c]
Toyota	(I)	694,296	38.5	2,209,123.636
VW	(LT)	40,067	20.7	237,111.473
VW	(I)	282,789	28.8	1,202,835.156
		14,006,358		62,632,528.233

[a]Following NHTSA's CAFE reports, we present separate estimates and data for each manufacturer's fleet based on its division according to whether its products are (D) domestic in origin, (I) import in origin, or (LT) considered to be part of the light truck fleet (including pick up trucks, SUVs, minivans, and cross-over vehicles)

[b]Shaded vehicle indicates that the manufacturer is subject to a fine under CAFE

[c]Fleet tons of carbon dioxide is an estimate of the tons of CO_2 produced by each manufacturer's fleet. It is derived by multiplying the number of vehicles sold by 12,500 miles (the estimated annual miles driven by an average passenger vehicle), dividing by the average EPA-fleet-estimated miles per gallon. This yields an estimate for the number of gallons of gasoline each fleet burns. The estimated gallons burned are multiplied by 19.6 (the number of pounds of carbon dioxide produced when a gallon of gasoline is burned) to produce the number of pounds of carbon dioxide each fleet of vehicles produces. This figure is divided by 2,000 to arrive at the tons of carbon dioxide each fleet is estimated to produce

27.5 mpg for passenger cars and the 22.5 mpg for light trucks, allowing for within-manufacturer credit trading; e.g., applying a manufacturer's passenger car CAFE credits to its light truck deficit). These data indicate that CAFE penalties apply to only a very small percentage of the vehicle fleet each year. As a result, it is unlikely that CAFE penalties as currently designed will have much influence on the production of global warming gases.

To address this issue further, we estimated the number of standard tons of carbon dioxide each manufacturer's fleet would produce (Table 12.1, column 4) when driven an average of 12,500 miles (Table 12.1, column 4). Across all manufacturers, this fleet of vehicles would produce more than 62.6 million standard tons of carbon dioxide in a year. Since CAFE targets the most inefficient vehicles, we expect that manufacturers subject to CAFE fines in this year would produce a larger share of carbon dioxide than the percentage their fleets represent of the entire US vehicle fleet sold that year (4.46%). Surprisingly, these violators produce only 4.80% (or 3,009,045.485 tons) of the total fleet's carbon dioxide output, or only slightly more than their share of the market. This outcome indicates that CAFE can only be expected to have a minor impact on carbon dioxide outputs.

Our initial examination indicates that the vast majority of vehicles sold in the United States in 2008—95.54%—met CAFE requirements either directly, or through allowable within-manufacturer credit trading (either on an annual basis or using the 3-year credit banking provisions). To be sure, CAFE has, in the long run, forced manufacturers to improve vehicle efficiency. For example, when CAFE was initiated (1978), it set a standard of only 18 miles per gallon for passenger vehicles, which increased to 27.5 miles per gallon in 1990. At the same time since these standards have not been changed for passenger cars since 1990, and have increased by only 12.5% for light trucks since 1990, the impact of CAFE regulations on the production of carbon dioxide by the 2008 US vehicle fleet was probably minimal. In our view, this outcome identifies the primary drawback of CAFE as a response to reducing the impact of the US vehicle fleet on the production of global warming gases. How can this shortcoming be addressed? As we suggest below, this requires re-conceptualizing CAFE fines so that they more directly address the production of global warming gases.

In place of CAFE fines based on vehicle fuel efficiency, we suggest a fee assessment model that applies to consumers and to all vehicles—what we call the Vehicle-Based CARbon Tax (V-CART) Fee. That is, every consumer who purchases a vehicle is assessed a climate change fee based on the vehicle's predicted carbon dioxide output calculated in relation to a vehicle's fuel efficiency and average number of miles driven by the typical American (which we set as 12,500 for purposes of the remaining discussion).

Why make the tax overt and assess individuals a vehicle-related climate change fee? Under CAFE individuals pay penalties when manufacturers raise the cost of the vehicle to offset the cost of the fee. Thus, making this fee transparent through V-CART has an educative value, especially if the fee is printed out and included on the vehicle pricing information. In short, it demonstrates the ecological impact of purchasing a vehicle. Each consumer will be exposed to the climate impact fee they must pay to purchase a specific vehicle, as well as the estimated number of tons of carbon dioxide that vehicle is calculated to produce in a year of driving. This information helps to uncover the harm associated with transportation and may impact how consumers use their vehicles. Second, an individual fee is more likely than a manufacturer-based fee to impact consumer choices. While consumers pay the fee or penalty under V-CART or CAFE, the V-CART fee is assessed directly to the consumer. Under existing CAFE regulations, manufacturers are fined directly and consumers only indirectly when they pay an increased price for their vehicle. The consumer, however, is generally unaware that the manufacturer has included this civil fine assessment in the price of the vehicles they sell. Third, we also selected an individual fee for all vehicle purchasers since all vehicles contribute to the production of global warming gases and the problem of climate change. Such a uniform tax or assessment is, in our view, a much fairer procedure than taxing only the most egregious violators. Fourth, by applying the tax to all purchases and hence all vehicles purchased, the assessment makes it clear that the use of any vehicle, not specific vehicles, contributes to climate change.

We should note that we do not envision V-CART as a penalty, but rather as the beginning of a broader mechanism designed to address the problem of global warming, or as part of a larger ameliorative process that promotes the environmental health of the planet. In our view, the fee collected under this program should be employed by the government to purchase carbon offsets from the global carbon offset market. These offset fees are paid by the government to carbon offset providers who purchase carbon offsets from companies that reduce their carbon outputs (e.g., through the use of alternative energy sources), or through other mechanisms such as the planting of trees. The next section addresses the calculation and implications of a V-CART.

Calculating Vehicle Carbon Tax (V-CART). In our model, each vehicle purchaser would be assessed a climate change assessment fee based on the carbon dioxide estimated to be produced by that vehicle. That estimate is based on the fuel efficiency of the vehicle purchased. Vehicle efficiency determines the environmental impact of each vehicle, on average, through its use. That is, as a vehicle is used, it consumes gasoline (or for electric models, the Environmental Protection Agency derived equivalent). In turn, the consumption of gasoline produces carbon dioxide. Burning one gallon of gasoline (assuming complete combustion) produces 19.6 lb of carbon dioxide. Thus, using the average miles per gallon obtained by a vehicle, an estimate that each vehicle travels an annual average of 12,500 miles, and knowing that burning one gallon of gasoline produces 19.6 lb of carbon dioxide, we can calculate the tons of carbon dioxide each vehicle is estimated to produce.

To calculate the V-CART, information on the cost of carbon offsets was needed. This was obtained by collecting data on the fee charged on by carbon offset providers in the United States using the listing maintained by the *Carbon Catalog* (www.carboncatalog.org/providers). From that list, we identified 26 carbon offset providers in the United States with a known or listed carbon offset fee. The average price charged by these providers for a one-metric ton carbon offset was $14.27. Since a metric ton is measured as 2,205 lb, and our carbon dioxide calculations employed standard US tons (2,000 lb), we adjusted this carbon offset average price to reflect its standard ton price, which we set to $12.94 to determine V-CART.

Since V-CART would be paid on the actual environmental impact of a specific vehicle, we calculated a number of V-CART examples for individual vehicles for the 2007 model year since this is the last year for which all fuel efficient data were available. These data are displayed in Table 12.2, and includes a cross-section of vehicles and types, but includes all Honda models for that year. Honda was selected as a complete example because it is widely regarded as producing the most efficient low-emission vehicles.

In order to summarize the fees associated with the proposed V-CART, Appendix presents the fee schedule for vehicles traveling an average of 12,500 miles each year by vehicle-rated miles per gallon. For vehicles in our 2007 sample, the Environmental Protection Agency estimated combined mileage ranges from 12 to 46 miles per gallon. A vehicle with a miles per gallon rating of 12 would be assessed a V-CART of $113.10, while a vehicle at the upper end of the range (46 mpg) would be assessed a V-CART of $34.46. These fees are relatively minor, and it is unlikely that, for example, the $78.64 difference between the most- and least-efficient vehicles on this list would act as a deterrent for consumers contemplating the purchase of an inefficient vehicle. We examine this point further in the discussion of these data.

Table 12.2 Examples of vehicle-specific carbon dioxide offset tax (V-CART), 2007 Model Year, combine EPA estimate for city–highway mileage[a]

Model	Engine	Combined EPA mpg	V-CART tax	Mid-price	V-CART%
BMW					
328i	3.0 A6	22	72.05	30,125	0.239
760Li	6.0 A12	15	105.68	73,675	0.143
X-5	4.8 A8	15	105.68	47,550	0.222
Cadillac					
CTS	2.8 M6	19	83.43	45,600	0.183
Ecalade (4)	6.2 A8	14	113.23	56,800	0.200
Chevrolet					
Aveo	1.6 M4	27	58.71	13,327	0.440
Cobalt	2.4 A4	24	66.05	17,575	0.376
Corvette	7.0 M8	18	80.06	73,950	0.108
Impala	3.5 A6	22	72.05	23,210	0.310
Impala	5.3 A8	18	80.06	26,450	0.303
Suburban (4)	6.0 A8	14	113.23	45,342	0.250
Trailblazer	4.2 A6	16	99.07	32,260	0.307
Ferrari					
599 GTB	5.9 A12	12	132.10	301,000	0.044
Ford					
Focus	2.0 A4	26	60.97	15,337	0.398
Mustang	4.0 M6	20	79.26	19,229	0.412
Mustang	5.4 M8	16	99.07	29,995	0.330
Explorer	4.6 A8	15	105.68	30,203	0.350
Expedition	5.4 A8	15	105.68	37,020	0.286
Escape-H	2.3 A4	30	52.84	28,112	0.188
Honda					
Accord	2.4 A4	25	63.41	21,140	0.300
Accord	2.4 M4	26	60.97	20,770	0.294
Accord	3.0 A6	21	75.48	27,750	0.272
Accord	3.0 M6	21	75.48	27,450	0.275
Accord-H[b]	3.0 H6	27	58.71	29,965	0.196
Civic	1.8 A4	29	54.66	15,825	0.345
Civic	1.8 M4	29	54.66	15,320	0.357
Civic	2.0 M4	23	68.92	24,777	0.278
Civic-H	1.3 H4	42	37.74	22,905	0.165
Fit	1.5 A4	30	52.83	14,666	0.360
Fit	1.5 M4	31	51.13	14,222	0.359
S2000	2.2 M4	20	79.26	31,860	0.249
Odyssey	3.5 A6	18	88.06	25,890	0.340
Ridgeline	3.5 A6	17	93.24	26,890	0.347
CR-V (2)[c]	2.4 A4	23	68.92	21,240	0.325
CR-V (4)[d]	2.4 A4	22	72.05	23,100	0.312
Element (2)	2.4 A4	22	72.05	19,420	0.371
Element (2)	2.4 M4	20	79.26	19,010	0.417

(continued)

Table 12.2 (continued)

Model	Engine	Combined EPA mpg	V-CART tax	Mid-price	V-CART%
Element (4)	2.4 A4	21	75.48	23,850	0.317
Element (4)	2.4 M4	20	79.26	24,656	0.321
Pilot (2)	3.5 A6	18	88.06	31,940	0.276
Pilot (4)	3.5 A6	17	93.24	33,250	0.280
Infiniti					
G35	3.5 M6	20	79.26	33,870	0.234
FX35	3.5 A6	18	88.06	40,080	0.220
Mercedes-Benz					
SLR	5.4 A8	13	121.94	471,000	0.026
R 63 AMG	6.2 A8	12	132.10	43,211	0.306
Mini (BMW)					
Cooper	1.6 M6	31	51.13	23,685	0.216
SAAB					
9-3	2.0 M4	22	72.05	27,612	0.261
97-X (4)	5.3 A8	16	99.07	39,620	0.250
Toyota					
Camry	2.4 A4	24	66.05	16,888	0.391
Camry	3.5 A6	23	68.92	22,730	0.303
Camry-H	2.4 H4	34	46.62	23,790	0.196
Corolla	1.8 A4	29	54.66	15,195	0.360
Corolla	1.8 M4	31	51.13	14,890	0.343
Prius-H	1.5 H4	46	34.46	21,750	0.158
Land Cruiser	4.7 A8	13	121.94	55,250	0.220
Volkswagen					
Eos	2.0 M4	24	66.05	28,790	0.229
Jetta	2.5 A4	22	72.05	16,250	0.443
Passat	2.0 M4	24	66.05	26,400	0.251
Touareg	4.2 A8	14	113.23	40,200	0.282
Volvo					
S40	2.5 A5	21	75.48	28,250	0.261
S80 (4)	4.4 A8	18	88.06	35,750	0.256
V70	2.4 A5	21	75.48	34,890	0.216
XC70	2.5 A5	19	83.43	34,660	0.241

[a]We used 2007 model years for this table because this is the last year that complete EPA data are publicly available from EPA's fuel economy Web site. Mid-price estimates were based on manufacturer's suggested retail price ranges for each vehicle, adjusting for automatic transmissions, engine upgrades, and four- or all-wheel drive

[b]Hybrid model

[c]Two-wheel drive model

[d]Four-wheel drive model or all-wheel drive model

Advantages of V-CART over CAFE

The V-CART has three primary goals. First, to provide a mechanism for alleviating environmental damage caused by the use of motor vehicles by assessing a motor vehicle tax on consumers rather than producers tied directly to a vehicle's rated efficiency and its estimated carbon dioxide output. To do so, the V-CART employs the free market price of a carbon dioxide sufficient to offset or cancel a vehicle's carbon dioxide pollution. Second, unlike existing responses, V-CART applies to all vehicles, not simply to the least efficient. Third, money collected under the V-CART program is used to offset vehicle-based carbon pollution. We address each issue briefly below.

Since its inception CAFE has prevented a large quantity of global warming gases from entering the environment. For example, for new requirements NHTSA (2008) estimates savings of 50.6 billion gallons of gasoline between 2020 and 2060, an average of 1.265 million gallons of gasoline or 12.4 million tons of carbon dioxide annually. In comparison, we estimate that the V-CART model proposed here would produce an additional 20% reduction in carbon dioxide emissions annually. Thus, while CAFE has a positive global warming prevention impact, it is not as efficient as other alternatives, nor does it produce the kinds of larger, more immediate reductions scientists suggest are needed to impact climate change (Lovelock 2006; Pearce 2008).

Second, the V-CART fee is directly tied to carbon dioxide pollution estimated to be emitted by a specific vehicle. When operated, all petroleum-based vehicles release global warming gases that contribute to climate change. Thus, fees associated with vehicle use should apply not just to those vehicles that pose the greatest threat of environmental damage, but to all vehicles in proportion to the environmental damage they cause. It strikes us as inequitable to penalize some car buyers simply because they purchase more damaging vehicles since all vehicles damage the environment and contribute to the production of global warming gases. Moreover, penalizing some car buyer/operators is an inefficient means for making the large reductions in global warming gases necessary to address the problem of climate change.

Third, V-CART directly addresses the problem of climate change by stimulating the development of carbon neutral energy sources in all spheres of life by requiring V-CART fees to be used by the government to buy carbon offsets on the open market in amounts equivalent to the climate change impacts associated with operating motor vehicles. In this scenario, each consumer pays the cost associated with reducing their carbon footprint with the government acting as an intermediary in the process, by setting and collecting V-CART and employing those fees to buy carbon offsets.

In our view, there are several important advantages to this revised regulatory process. First, this regulatory approach makes each motor vehicle purchaser responsible for the carbon footprint attributed to their use of a motor vehicle. Even when a manufacturer offers to pay the V-CART fee to induce a purchase, the consumer is still aware of the fee. Thus, the fee serves an educative function and makes consumers more aware of how the choices they make affect the environment. Second, V-CART creates a mechanism through which all motor vehicle purchasers "zero-out" their motor vehicle-related carbon contribution. Third, because V-CART directly regulates the flow of carbon in the environment it is a more efficient form of regulation than current CAFE rules, which regulated carbon dioxide outputs indirectly.

We are not suggesting that the government should ignore vehicle efficiency, or that CAFE fuel economy standards be eliminated. Rather, what we are suggesting is a supplemental form of regulation which has a more immediate impact on the environment than more narrowly targeted goals of fuel efficiency contained within CAFE. To be sure, given the diminishing quantity of oil left in the world's oil reserve, it remains necessary to address vehicle fuel efficiency, especially given the level of petroleum use in the United States, a nation that consumes approximately one-quarter of the world's petroleum products annually, and which contributes nearly one-fifth of annual global warming gas pollution (Carbon Dioxide Information Analysis Center 2009).

Economists sometimes worry that taxes and assessments fees, of the type we propose, are regressive—that is, they have an unequal distribution that disadvantages those in lower-income brackets. This tendency may appear evident in the V-CART when examining Table 12.2, which displays data on the V-CAT to vehicle cost ratios (the last column in that table). At the upper price range, V-CART costs are indeed small compared to the purchase price of inefficient vehicles, which tend to also be luxury oriented or exotic, suggesting a regressive tax effect. On average the V-CART is less than 0.45% of vehicle cost, or $45/$10,000 in vehicle price, and the distribution of V-CART over the entire range of vehicles is between $38 and $132. Thus while somewhat regressive in relation to the potential income of purchasers, the point of V-CART is to eliminate rather than redistribute each consumer's effect on the environment. A redistribution effect that takes income into account could be accommodated quite simply by proportionately increasing V-CART fees on carbon emissions as a vehicle's emission rate rises.

Specific Limitations of V-CART

The V-CART solution is not without its limitations. For instance, we have not explored whether it is wise to tax all vehicles, such as very efficient hybrids. Indeed, one could argue that it would make sense to use an additional tax, such as the gas guzzler tax (see, Energy Tax Act, 1978, Pub.L. 95-618, 92 Stat. 3174) to not only fine those who purchase inefficient vehicles but to provide tax credits for those who purchase efficient hybrids (currently, not all hybrid are "efficient" and return gas mileage ratings similar to traditionally fueled vehicles). Again, this may be problematic to the extent that it is regressive because hybrids tend to be more expensive to purchase (see Brisman 2009). Moreover, Brisman (2009) suggests that the consumption of hybrid vehicles may actually reinforce class boundaries. Currently, the gas guzzler tax applies only to passenger cars weighting less than 6,000 lb and does not apply to light trucks, including SUVs.

In addition, we do not possess sufficient information to determine if the carbon offset market currently has access to a sufficient quantity of offsets to cancel out the carbon dioxide contributions of the entire fleet of new US automobiles. If a sufficient level of carbon offsets is not available, then clearly this limits the usefulness of our alternative, though it does not necessarily imply that a more limited version of this alternative should not be employed. Any regulatory plan that would reduce the US's contribution to climate change is welcome, since the United States is the world's largest producer of global warming gases. Finally, in our discussion of the V-CART we have modeled our approach somewhat after existing CAFE penalties which apply only to new vehicle sales. Clearly, vehicles continue to contribute to the production of global warming gases throughout their life time of use. Thus, there may be some need to extent this regulation to all registered automobiles, or to develop an alternative regulatory mechanism for vehicles in use. Again, whether there are sufficient offset credits available in the marketplace at the current time to allow such a form of regulation is beyond our knowledge. Nevertheless, such a plan would stimulate demand for carbon offsets, and encourage the more rapid development of programs and technologies that provide carbon offsets.

During preparation of this manuscript, President Obama announced an agreement between the Environmental Protection Agency, the National Highway Transportation and Safety Administration, and United States automakers on a new CAFE procedure and standards. The new CAFE regulations, to take effect in 2016, mark a step in the direction suggested by our research. Specifically, by 2016, CAFE will regulate both vehicle miles per gallon and carbon dioxide tailpipe emissions. In 2016, a new carbon dioxide tailpipe emission standard will take effect, 250 g/mile. To better comprehend this standard, it is useful to calculate the relationship between the tailpipe emission standard and the 2016 CAFE mileage requirement of 39 miles per gallon (for passenger cars). A vehicle that meets the latter efficiency requirement will burn 1 gallon of gasoline, producing 19.6 lb of carbon dioxide, or 8,890.56 g of carbon dioxide. Per mile, the actual tailpipe emission for this vehicle would be 227.96 g (8,890.56/39). Thus, the proposed 2016 carbon dioxide tailpipe emission standard is about 9.7% higher than the expected carbon dioxide emissions produced by a vehicle that would obtain a 39 mile per gallon rating given perfect combustion and a fuel mixture without other chemical additives. Since fuel combustion in a typical gasoline engine is not expected to be complete, and since gasoline includes additives, it is likely that burning one gallon of gasoline in an average vehicle produces less than the 227.96 g of carbon dioxide from the complete combustion of 1 gallon of pure, additive-free octane. Thus, the proposed carbon dioxide standard is not as strict as the 2016 CAFE mileage standard.

Moreover, while the newly proposed tailpipe emission standard is a step in the right direction since it directly regulates carbon dioxide emissions, it fails, in our opinion, to go far enough. In contrast to our proposal which includes a V-CART assessed on all new vehicles, the 2016 proposal would continue the traditional CAFE practice of fining only non-complaint manufacturers. As noted in our analysis of this issue, few manufacturers are in noncompliance. But more importantly, all vehicles contribute to the production of global warming gases, and given the severity of the current trend in global warming gas production, broader regulations that offset vehicle contributions to climate change are necessary.

Conclusion

It is important that green criminology address the kinds of policies that are consistent with examining the control of environmental harms (see Chaps. 7 and 6). This chapter constitutes an example of one way in which green criminologists can examine relevant policy matters using global warming policies related to automobile carbon dioxide pollution. Our example illustrates what is essentially a tax on automobile users. This small tax can be used to offset automobile global warming gas production by devoting the collected revenue toward the purchase of carbon offsets and by investing those revenues in non-fossil fuel technologies. While short term, this remains an ambitious goal given the economic and political power wielded by automakers and oil companies in the US.

In closing we wish to point out the broad limitations of the policy we described above. Without a doubt, our recommendation could be implemented in the short term to address global warming gas producing by automobiles in the United States. Our automobile pollution control policy addresses only one of several potential sustainability concerns related to automobile use—namely, climate sustainability associated with vehicle global warming gas production. Our policy does not address related matters such as the sustainability of the fossil fuel resources needed to operate vehicles at this point in history. Nor does it address sustainability issues related to the production of automobiles or automotive parts and supplies. One of the important sustainability issues also omitted from consideration which has global warming implications is the building and maintenance of roadways. Roadways are made from tar or concrete, and the extraction and manufacture of both of these substances has large global warming consequences. Likewise, the building of roadways involves destruction of natural environments, which clearly poses a threat to the sustainability of eco-systems and poses related environmental threats from hazardous chemical run-off from roadways.

Our policy also does not directly address a key issue with respect to automobiles—fuel economy. Because of our interest in generating a policy that can be implemented in the short term that would counteract the more immediate impacts of vehicles on the environment, we have not addressed vehicle fuel efficiency and technology. Certainly such policies would make useful adjuncts to the perspective taken here. Long-term policies, however, are more difficult to construct. One reason for this observation relates to the fossil fuel treadmill of production which includes very powerful industries such as automakers and oil, coal, and natural gas, as well as railroads which ship large quantities of products used to fuel the modern automobile culture in the United States (Gould et al. 2008). In large part, it is that automobile culture itself that must be addressed by long-term environmental sustainability policies. Besides a small number of cities in the Northeast (New York, New York; Jersey City, New Jersey; and Newark, New Jersey), most cities are car dependent, and the large majority of residents in cities other than the three mentioned above (more than 70% on average) commute to work by car. American society is largely organized around the car transportation, and until that challenge is addressed by long-term policy, America's environmental footprint will remain large and cause extensive environmental damage that impact people across the globe. In short, our policy does nothing to address the importance of mass transportation as one of the most effective carbon reduction methods and therefore will only legitimate individual transportation if long-term policy does not promote mass transportation.

As radical environmentalist John Bellamy Foster (2002, p. 19) notes, one of the largest and most difficult issues to address in the intersection of carbon-based economies and capitalism. Fossil fuels store significant, concentrated energy, and drive the machinery of capitalism. These fuels have generated large markets and large corporations devoted to bringing those commodities to market. As a result, the extraordinary level of profit available in this industry has lead "capital to structure the energy economy around fossil fuels." This fossil fuel economy will not disappear overnight, and small steps toward dismantling our domination by fossil fuels will be required to protect the long-term health of the environment. The development of legal mandates and harm associated with global warming suggest that green criminologists should be engaged in these efforts.

Appendix: V-CART for Vehicles Traveling 12,500 Miles Annually, at the May 2009 Estimated Carbon Offset Fee of $12.94 per Standard Ton

mpg	V_CART
12	132.10
13	121.94
14	113.23
15	105.68
16	99.07
17	93.24
18	88.06
19	83.43
20	79.26
21	75.48
22	72.05

(continued)

(continued)

mpg	V_CART
23	68.92
24	66.05
25	63.41
26	60.97
27	58.71
28	56.61
29	54.66
30	52.84
31	51.13
34	46.62
35	45.29
40	39.63
42	37.74
45	35.23
46	34.46
50	31.70
60	26.42

References

Beirne, P. (1999). For a nonspeciesist criminology: Animal abuse as an object of study. *Criminology, 37*, 117–148.

Brisman, A. (2009). It takes green to be green: Environmental elitism, "ritual displays", and conspicuous non-consumption. *North Dakota Law Review, 85*, 329–370.

Brumbelow, M. L., Teoh, E. R., Zuby, D. S., & McCartt, A. T. (2008). Roof strength and injury risk in rollover crashes. Arlington, VA: Insurance Institute for Highway Safety. Online. http://www.iihs.org/research/topics/pdf/r1098.pdf. Accessed 2 Aug 2011.

Burns, R. G., & Lynch, M. J. (2002). Another fine mess...a preliminary examination of the use of fines by the National Highway Traffic Safety Administration. *Criminal Justice Review, 27*, 1–25.

Burns, R. G., & Lynch, M. J. (2004). *Environmental crime: A sourcebook*. New York: LFB Scholarly Publishing.

Burns, R. G., Lynch, M. J., & Stretesky, P. B. (2008). *Environmental law, crime and justice*. New York: LFB Scholarly Publishing.

Carbon Dioxide Information Analysis Center. (2009). Preliminary 2006–2007 global and national estimates. Online. http://cdiac.ornl.gov/trends/emis/meth_reg.html. Accessed 2 Aug 2011.

Crandall, R. W. (1985). Why should we regulate fuel-economy at all? *Brookings Review, 3*, 3–7.

Crandall, R. W. (1990). The changing rational for motor-vehicle fuel-economy regulations. *Regulation, 13*, 10–12.

Crandall, R. W., & Graham, J. (1989). The effects of fuel economy standards on automobile safety. *Journal of Law and Economics, 32*, 97–119.

Crandall, R. W., Gruenspecht, H. K., Keeler, T. E., & Lave, L. B. (1986). *Regulating the automobile*. Washington, DC: The Brookings Institute.

DeCicco, J., & Fung, F. (2006). Global warming on the road: The climate impact of America's automobiles. Online. http://www.edf.org/documents/5301_Globalwarmingontheroad.pdf. Accessed 2 Aug 2011.
Environmental Defense Fund. (2007). Tallying greenhouse gases from cars: A new look at ways we can downshift heat trapping emissions. Online. http://www.edf.org/article.cfm?ContentID=5300. Accessed 7 Sept 2010.
Environmental Protection Agency. (2010a). EPA and NHTSA propose first-ever program to reduce greenhouse gas emissions and improve fuel efficiency of medium- and heavy-duty vehicles: regulatory announcement. No. EPA-420-F-10-901. Online. http://www.nhtsa.gov/staticfiles/rulemaking/pdf/cafe/CAFE_2014-18_Trucks_FactSheet-v1.pdf. Accessed 2 August 2011.
Environmental Protection Agency. (2010b). EPA and NHTSA finalize historic national program to reduce greenhouse gases and improve fuel economy for cars and trucks. No. EPA-420-F-10-014. Online. http://www.nhtsa.gov/staticfiles/rulemaking/pdf/cafe/CAFE_2014-18_Trucks_FactSheet-v1.pdf. Accessed 2 Aug 2011.
Environmental Protection Agency. (2010c). Light-duty vehicle greenhouse gas emission standards and corporate average fuel economy standards; final rule. *Federal Register, 75*(88), 25325–25728.
Foster, J. B. (2002). *Ecology against capitalism*. New York: Monthly Review Press.
Goldberg, P. K. (1998). *The effects of corporate average fuel efficiency standards in the US* (pp. 1–33). XLVI: Journal of Industrial Economics.
Gould, K. A., Pellow, D. N., & Schnaiberg, A. (2008). *The treadmill of production: Injustice and unsustainability in the global economy*. Boulder, CO: Paradigm Publishers.
Gramlich, J. (2009). Gas prices and fuel efficiency in the U.S. automobile industry: Policy implications of endogenous product choices. Doctoral dissertation. Yale University.
Greene, D.L. (1997). Why cafe worked. Online. <http://ntl.bts.gov/lib/000/700/725/cafeornl.pdf. Accessed 2 Aug 2011.
Heavenrich, R. M., & Hellman, K. H. (1996). *Light-duty automotive technology and fuel economy trends through 1996*. Ann Arbor, MI: US EPA.
Insurance Institute for Highway Safety. (2003). In real world crashes, airbags with head protection saving lives, news release. Online. http://www.iihs.org/news/2003/iihs_news_082603.pdf. Accessed 2 Aug 2011.
Kahn, M. E. (1996). New evidence on trends in vehicle emissions. *RAND Journal of Economics, 27*, 183–196.
Lovelock, J. E. (2006). *The revenge of Gaia: Earth's climate crisis of the fate of humanity*. New York: Basic Books.
Lynch, M. J. (2007). *Big prisons, big dreams: Crime and the failure of the US prison system*. New Brunswick, NJ: Rutgers University Press.
Lynch, M. J., Burns, R., & Stretesky, P. B. (2010). Global warming and state-crime: The politicization of global warming under the Bush administration. *Crime, Law, and Social Change, 54*(3–4), 213–239.
Lynch, M. J., & Stretesky, P. B. (2010). Global warming, global crime: A green criminological analysis. In R. White (Ed.), *Global environmental harm and the greening of criminology*. Devon: Willan Publishing.
National Academy of Sciences. (2003). *Effectiveness and impact of corporate average fuel economy (CAFE) standards*. Washington, DC: National Academy Press.
National Highway Traffic Safety Administration. (2011a). Summary of fines. Online. http://www.nhtsa.gov/staticfiles/rulemaking/pdf/cafe/Fines_Collected_062010.pdf. Accessed 2 Aug 2011.
National Highway Traffic Safety Administration. (2011b). CAFE overview – Frequently asked questions. Online. http://www.nhtsa.gov/cars/rules/CAFE/overview.htm. Accessed 2 Aug 2011.
National Highway Traffic Safety Administration. (1997). *Relationship of vehicle weight to fatality and injury risk in model year 1985–1993 passenger cars and light trucks*. Washington, DC: US Department of Transportation.

National Highway Traffic Safety Administration. (2008). *Final environmental impact statement: Corporate average fuel economy standards, passenger cars and light trucks, model years 2011–2015*. Washington, DC: National Highway Traffic Safety Administration.

Pearce, F. (2008). *With speed and violence: Why scientists fear tipping points in climate change*. Boston, MA: Beacon.

Stretesky, P. B., & Lynch, M. J. (2009). A cross-national study of the association between per capita carbon dioxide emissions and exports to the United States. *Social Science Research, 38*, 239–250.

Van Biesebroeck, J., & Leuven, K. U. (2010). The demand for and the supply of fuel efficiency in models of industrial organization. Paper Presented at the International Transportation Forum, Paris, France. Online. http://www.internationaltransportforum.org/jtrc/DiscussionPapers/DP201009.pdf. Accessed 2 Aug 2011.

White, R. (2010a). Introduction. In R. White (Ed.), *Global environmental harm: Criminological perspectives*. Devon: Willan.

White, R. (2010b). Globalisation and environmental harm. In R. White (Ed.), *Global environmental harm: Criminological perspectives*. Devon: Willan.

White, R. (2011). *Transnational environmental crime: Toward an eco-global criminology*. London: Routledge.

Zakrzewski, S. F. (2002). *Environmental toxicology*. New York: Oxford University Press.

Index

R. White (ed.), Climate Change from a Criminological Perspective,
DOI 10.1007/978-1-4614-3640-9, © Springer Science+Business Media New York 2012

O

P

R

MIX
Papier aus verantwortungsvollen Quellen
Paper from responsible sources
FSC® C105338

If you have any concerns about our products,
you can contact us on
ProductSafety@springernature.com

In case Publisher is established outside the EU,
the EU authorized representative is:
Springer Nature Customer Service Center GmbH
Europaplatz 3, 69115 Heidelberg, Germany

Printed by Libri Plureos GmbH
in Hamburg, Germany